SpringerBriefs in Applied Sciences and Technology

SpringerBriefs present concise summaries of cutting-edge research and practical applications across a wide spectrum of fields. Featuring compact volumes of 50 to 125 pages, the series covers a range of content from professional to academic.

Typical publications can be:

- A timely report of state-of-the art methods
- An introduction to or a manual for the application of mathematical or computer techniques
- A bridge between new research results, as published in journal articles
- A snapshot of a hot or emerging topic
- An in-depth case study
- A presentation of core concepts that students must understand in order to make independent contributions

SpringerBriefs are characterized by fast, global electronic dissemination, standard publishing contracts, standardized manuscript preparation and formatting guidelines, and expedited production schedules.

On the one hand, **SpringerBriefs in Applied Sciences and Technology** are devoted to the publication of fundamentals and applications within the different classical engineering disciplines as well as in interdisciplinary fields that recently emerged between these areas. On the other hand, as the boundary separating fundamental research and applied technology is more and more dissolving, this series is particularly open to trans-disciplinary topics between fundamental science and engineering.

Indexed by EI-Compendex, SCOPUS and Springerlink.

More information about this series at http://www.springer.com/series/8884

Patrick Rérat

Cycling to Work

An Analysis of the Practice of Utility Cycling

 Springer

Patrick Rérat ⓘ
Institute of Geography and Sustainability
and Observatory for Cycling and Active
Mobilities
University of Lausanne
Lausanne, Switzerland

Translation by Hannah Juby and Becky Warner of Express Language

ISSN 2191-530X   ISSN 2191-5318 (electronic)
SpringerBriefs in Applied Sciences and Technology
ISBN 978-3-030-62255-8   ISBN 978-3-030-62256-5 (eBook)
https://doi.org/10.1007/978-3-030-62256-5

Acknowledgements

This work was supported by the University of Lausanne, the Canton of Vaud and Romande Energie as a part of the Volteface research programme. I am grateful for this support.

I would like to address a special word of thanks to Gianluigi Giacomel and Antonio Martin for their work on the statistical analysis of the large-scale survey on which this book is based. I would also like to thank the other previous and current members of my research team in the geography of mobilities at the University of Lausanne. Their involvement in various research and teaching projects on mobilities and on cycling have enriched the analysis and reflections presented in this book.

A first version of the book was published in French by the Editions Alphil—Presses universitaires suisses. The English version has been reworked and extended for an international readership. I would like to thank Hannah Juby and Becky Warner of Express Language for having translated the manuscript.

Finally, the research would not have been successful without the support of PRO VELO Switzerland, who agreed to send the survey to the participants in the *bike to work* scheme. Almost 14,000 of them took the time to fill in a (long) questionnaire! Particular thanks are due to them.

Lausanne, Switzerland   Patrick Rérat
August 2020

Contents

About the Author

Patrick Rérat is Full Professor in geography of mobilities at the University of Lausanne, Switzerland. He holds a PhD in social sciences from the University of Neuchâtel and he has been a visiting researcher at King's College London, HafenCity University Hamburg and Loughborough University.

His research interests focus on urban planning and sustainable development, which he addresses through the various forms of residential and everyday mobilities. He has published about 50 papers in journals such as Transportation Research Part A, Urban Studies, Transactions of the Institute of British Geographers, International Journal of Housing Policy, Applied Mobilies, Environment and Planning A, etc.

He is the co-founder and co-director of the Observatory for Cycling and Active Mobilities ('OUVEMA'), launched in 2020 at the University of Lausanne.

List of Figures

List of Tables

Chapter 1
Introduction

In 1817, Karl von Drais travelled 14 kms around Mannheim on a strange two-wheeled vehicle, a 'running machine', which is considered to be the ancestor of the bicycle. Two centuries after its invention, the bicycle is back on centre stage, and it may just be one of the keys to the mobility of the future.

1.1 The Return of the Bicycle

What a journey the bicycle has been on! Baron von Drais's running machine marks the beginning of a series of innovations which gave rise to the bicycle in the late nineteenth century. Initially, a leisure pursuit for the bourgeoisie, the bicycle, thanks to mass production, became a cheap and popular means of transportation for workers. At the end of the Second World War, increased purchasing power and the proliferation of motorised transport caused the practice to collapse in all industrialised countries [9, 19, 20] including Switzerland [12, 21]. In the 1970s, an upswing was observed in the Netherlands and Denmark while, in general, the bicycle continued to lose relevance. Finally, over the last 15 years, a comeback has been observed in Western cities, with increasing numbers of urban centres promoting cycling. Construction of infrastructure, the emergence of new types of bicycles (electrically assisted bikes, bike share services, etc.) and a renewed image have allowed the number of users to increase. In central Copenhagen, bike traffic is now greater than car traffic and the same is now observed during rush hour in central London.

So, what's the situation like in Switzerland? At the national level, the growth in the proportion of travel completed by bicycle is modest and recent. In the larger cities, however, the increase is evident as we will see in Sect. 3.2. The bicycle has also become a political object. In 2018, about 75% of Swiss people agreed to incorporate the principle of promoting cycling—without any binding measures though—in the federal constitution. The city of Bern has launched a 'bike offensive' (*Velo-Offensive*)

P. Rérat, *Cycling to Work*,
SpringerBriefs in Applied Sciences and Technology,
https://doi.org/10.1007/978-3-030-62256-5_1

and aims to become the Swiss capital of cycling (*Velo-Hauptstadt*). Other urban centres—Basel, Winterthur, Lucerne, etc. —are not far behind, and contest this title. In Zurich, in 2017, signatures were collected in record time to support an initiative to challenge the authorities over a claimed delay in the provision of cycling facilities.

Little is known about cycling practices in Switzerland and there is little research on the subject. The imagery and conversations around cycling are at the very least contradictory, even clichéd. The bicycle is the vehicle of the penniless student; it is a means of transport for the young trendy executive, a toy for children, a flexible and rapid means of transport, an experience associated with holidays; cyclists play fast and loose with traffic rules; users have been left behind due to decades of planning policy prioritising cars, etc.

Given the recent return to grace of the bicycle in urban policy and the current challenges in the field of mobility, it is important to investigate the utilitarian dimension of cycling.[1] This is the objective of this research, which is based on a survey answered by nearly 14,000 *bike to work* participants. Each year, this action brings together people who agree to use bicycles as much as possible in their commuting journeys during May and/or June. The various empirical chapters of this book relate to the uses of cycling, commuter cyclists' access to means of travel, the skills required to manage daily journeys, the motivations for choosing to cycle, the barriers encountered, as well as user evaluations of traffic conditions, the quality of amenities and infrastructures, and the focus on cycling by the public authorities. Because of its scope, this approach captures in detail the various dimensions of utility cycling as well as its shortcomings in the case of Switzerland. By doing so it brings theoretical and empirical elements to the research on and the politics of cycling in the many countries where cycling culture is being redefined.

1.2   Why Should We Care About Cycling?

In a world described as being increasingly fast, fluid and (inter)connected, is not cycling an anachronism? What role can it play in the transportation system? What problems can it contribute to solve? What are the arguments put forward to promoting cycling?

A first challenge is energy transition. Lifestyles, travel habits, indeed the entire economic system, all operate on the basis of abundant and cheap energy. This organisation is being challenged due to climate change and the heavy dependence on

---

[1]Utility cycling emphasises its function as a means of transport, while recreational cycling refers to a leisure or sporting activity.

non-renewable resources. Energy transition involves objectives such as the progressive rejection of fossil fuels, the promotion of renewable resources, and a reduction in greenhouse gas emissions.[2]

But the energy transition simply cannot be implemented without another transition: the 'mobility transition' or the transition to 'low-carbon mobility' [6]. This involves a change in mobility practices, or at least in the way in which they are exercised. Mobility plays a central role in the energy issue. In Switzerland, transportation accounts for 36% of final energy consumption, and 94% of this share comes from fossil sources [15]. Greenhouse gas emissions, for their part, owe 32% to transportation, despite international air traffic not being counted [14]. Other environmental impacts, on both the local and regional scale, are also listed. They relate, in particular, to the emission of various pollutants and suspended particulate matter, which have significant impacts in terms of public health.

Three action levers, three verbs, summarise the discourse on mobility transition: improve, transfer and avoid [5]. Improving refers to technological solutions, which aim to reduce negative externalities by making the transport system more efficient and by opting, for example, for alternatives to petroleum fuels (for example, electricity). Transferring involves promoting more resource-efficient forms of mobility by favouring shared forms (public transport, carpooling and car sharing) and demotorised forms or active mobility.[3] Avoiding means encouraging lifestyles that are no longer based on high mobility but on a more restricted spatial scale and on the valorisation of proximity.

The mobility transition is therefore not just a technological issue; it is eminently social and political in nature and is intimately linked to the very organisation of lifestyles and the way in which cities and territories are organised. From this perspective, cycling can make a significant contribution. It only requires a low level of energy for both its manufacture and its use and it is characterised by the absence of pollutants and greenhouse gas emissions. The e-bike requires more materials, energy and a battery. However, it is characterised by much lower greenhouse gas emissions than other motorised vehicles [4].

Cycling has other equally significant benefits. In terms of public health, it facilitates the reintroduction of physical activity into increasingly sedentary lifestyles and reduces the problems that result therefrom. The studies agree on the health benefits of cycling: reduced risk of and mortality from stroke and infarction, reduced incidence of and mortality from certain cancers, prevention of diabetes and obesity, etc.[2, 8, 11]. For this reason, in 2017, nearly 500 doctors in Geneva called for the prioritisation of developing secure cycling facilities as part of the canton's political agenda. There are also positive effects for users of e-bikes, which increase their level of physical activity [7].

---

[2]The research presented in this book is taken from the Volteface research programme. Relating to the social challenges of the energy transition, a dozen of projects have been carried out at the University of Lausanne with the support of Romande Energie and the Canton of Vaud [13].

[3]Active mobility includes forms of travel that utilise human energy (walking, cycling, scooters, skateboarding, etc.). In Switzerland, we speak most frequently about 'soft mobility' in French (mobilité douce) or, in German, 'slow transport' (Langsamverkehr).

Overall, the research shows that the benefits of regular cycling outweigh the negative consequences of exposure to air pollution and the risk of accidents. According to the studies reviewed by Héran ([9], 163), 'Motorists breathe air which is twice as polluted as that of cyclists and four times more than that of pedestrians, with wide variations depending on pollutants and the routes travelled. These results are explained by the different distance of users from pollutants that stagnate at ground level. However, by exerting themselves, cyclists inhale 2.4 times more air than motorists, which slightly more than negates this advantage'. Air quality improves, however, as soon as cyclists are moved a short distance away from the flow of motor vehicles.

With regards to the risk of accidents, this should not be underestimated, but can be reduced considerably by means of adequate infrastructure. A so-called *safety in numbers* phenomenon is also observed: the more cyclists there are, the less they are proportionally victims of accidents [10]. This is explained by greater visibility of cyclists, greater attention by motorists, more cycling amenities, traffic calming measures, etc. The lowest accidentology rates (in proportion to number of cyclists and kilometres travelled) are thus observed in the countries of northern Europe.

An additional element is the growing emphasis on quality of life and conviviality, especially in cities where the negative externalities of car traffic are felt the most. Ecological, silent, and economical on space, bicycles are particularly attractive in the context of urban centres which are rethinking the role of cars due to their air and noise pollution. In Switzerland, one in seven people during the day, and one in eight people at night, are exposed to harmful or bothersome noise emissions from road traffic when at home. Exposure to noise from road traffic is most widespread in the heart of urban areas, where one-third of the population is affected [16]. The development of cycling would start a fundamental movement towards a relative calming of cities.

The bicycle is also interesting due to its smaller footprint, both in terms of travel space and parking space. The promotion of cycling is considered by some urban planners as an opportunity to rethink and transform circulation spaces into public spaces [1, 3]. The low spatial footprint of active mobilities has become crucial with the need to guarantee physical distancing during the COVID crisis. To cope with the diminished carrying capacity of transit, many cities have used 'tactical urbanism' and implemented pop-up bike lanes.

In economic terms, using a bicycle is less expensive than other modes of travel—with the exception of walking—not only for the users but also in terms of investments in infrastructure. In addition, because of their speed and flexibility, bicycles are effective on short journeys and, in particular, in urban areas where the volume of traffic and access or parking restrictions make car use less competitive. Mechanical bicycles[4] also, compared to walking, make it possible to travel three or four times further for the same amount of energy expended, i.e. to have access to a territory 10 to 15 times the size [9], 31).

The promotion of cycling mobility could seem to go against the grain in the increasingly mobile society before the COVID crisis. This assumption ignores the

---

[4]The terms mechanical, conventional or traditional bicycle in this book denote bicycles, which are propelled purely by the energy expended by their users (in contrast to e-bikes).

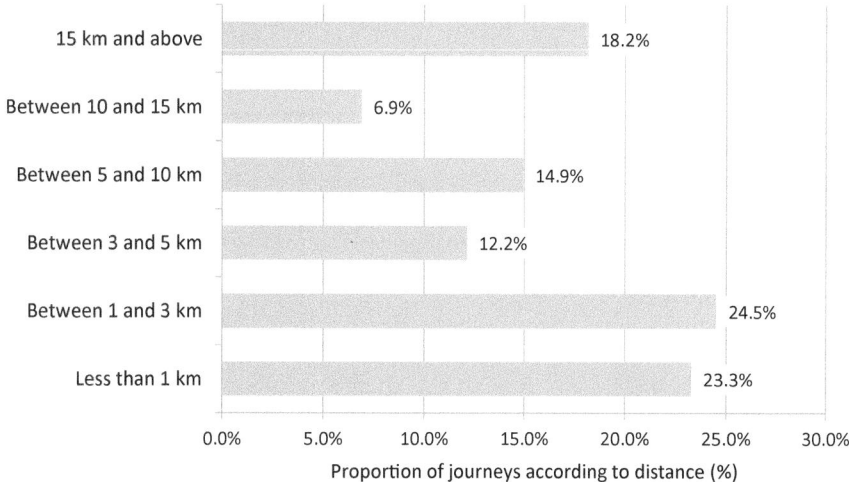

Fig. 1.1 Distribution of journeys by distance travelled, 2015 (*Source* Mobility and Transport Microcensus)

fact that many journeys took place over short distances. In Switzerland, according to the Mobility and Transport Microcensus,[5] 60% of journeys, irrespective of their reason, do not exceed 5 kms, a distance for which cycling is considered an attractive option [18] (Fig. 1.1). This proportion rises to 48% for journeys to the workplace.[6] Even over short distances, the share of journeys made by car is significant: 18.4% respectively for commuting journeys of less than 1 km, 46.1% for journeys of between 1 and 3 kms and 56.6% for those of between 3 and 5 kms (ibid.).[7] Of course, some of these journeys cannot be made using another mode of transport due, in particular, to topography, the physical condition of the travellers or the need to complete a succession of journeys.[8] It appears, however, that there is a substantial margin for progress in the promotion of cycling and the organisation of proximity, especially if we compare Switzerland and its cities with other contexts (see 3.2).

Cycling is, of course, not the only solution to mobility challenges. It cannot meet the transport needs for a certain number of uses, population groups and territorial contexts. However, it does seem pertinent to increase its place within the mobility

---

[5]This telephone survey is carried out every 5 years using a sample of more than 50,000 people. The latter are questioned in detail about their mobility behaviour on a specific reference day (the day before the survey). The survey takes place throughout the year to avoid seasonal bias.

[6]Another source, the Structural survey, shows lower figures though: 6.7% of the commuters (people working outside their home) travel less than 1 km, 25% between 1 and 5 kms and 21.3% between 5 and 10 kms. This would mean that half of the commuters work at a distance that is accessible by e-bike at least [17].

[7]Furthermore, the car occupancy rate is low: 1.10 person for commuting journeys, 1.56 for all journeys.

[8]However, only 27% of circuits (a circuit is a succession of journeys starting at home and returning back there) are made up of more than a simple round trip (OFS and ARE 2017).

ecosystem and planning policies. The development potential of the bicycle is all the more promising given the emergence and widespread nature of offerings such as e-bikes. Promoting cycling to a wider audience requires better knowledge of this practice. However, little information is available on those people who have already adopted the bicycle as a means of transport, on their motivations and on the barriers that they face. The next chapter proposes a grid for analysing bicycle usage, on which this research work is based.

References

1. S. Bendiks, A. Degros, *Cycle Infrastructure* (nai010 publishers, Rotterdam, 2013)
2. C.A. Celis-Morales, D.M. Lyall, P. Welsh, J. Anderson, L. Steell, Y. Guo, R. Maldonado, D.F. Mackay, J.P. Pell, N. Sattar, J.M.R. Gill, Association between active commuting and incident cardiovascular disease, cancer, and mortality: prospective cohort study. BMJ j1456 (2017) https://doi.org/10.1136/bmj.j1456
3. A. Degros, Traffic space is public space! Les espaces de trafic sont des espaces publics! GeoAgenda **1**, 18–21 (2018)
4. E. Fishman, C. Cherry, E-bikes in the mainstream: reviewing a decade of research. Transp. Rev. **36**(1), 72–91 (2016). https://doi.org/10.1080/01441647.2015.1069907
5. M. Givoni, Alternative pathways to low carbon mobility, in *Moving Towards Low Carbon Mobility*. ed. by M. Givoni, D. Banister (Edward Elgar, Cheltenham, 2013), pp. 209–230
6. M. Givoni, D. Banister (eds.), *Moving Towards Low Carbon Mobility* (Edward Elgar, Cheltenham, 2013)
7. B. Gojanovic, J. Welker, K. Iglesias, C. Daucourt, G. Gremion, Electric bicycles as a new active transportation modality to promote health. Med. Sci. Sports Exerc. **43**(11), 2204–2210 (2011). https://doi.org/10.1249/MSS.0b013e31821cbdc8
8. T. Götschi, J. Garrard, B. Giles-Corti, Cycling as a part of daily life: a review of health perspectives. Transp. Rev. **36**(1), 45–71 (2016). https://doi.org/10.1080/01441647.2015.105 7877
9. F. Héran, *Le retour de la bicyclette: une histoire des déplacements urbains en Europe, de 1817 à 2050* (La Découverte, Paris, 2014)
10. P.L. Jacobsen, Safety in numbers: More walkers and bicyclists, safer walking and bicycling. Inj. Prev. **9**(3), 205–209 (2003). https://doi.org/10.1136/ip.9.3.205
11. L. Mertens, S. Compernolle, B. Deforche, J.D. Mackenbach, J. Lakerveld, J. Brug, C. Roda, T. Feuillet, J.-M. Oppert, K. Glonti, H. Rutter, H. Bardos, I. De Bourdeaudhuij, D. Van Dyck, Built environmental correlates of cycling for transport across Europe. Health Place **44**, 35–42 (2017). https://doi.org/10.1016/j.healthplace.2017.01.007
12. B. Meyer, *Vorwärts Rückwärts: Zur Geschichte des Fahrradfahrens in der Schweiz* (Bautz, Nordhausen, 2014)
13. N. Niwa, B. Frund, *Volteface. La transition énergétique : un projet de société* (Editions d'en bas & Editions Charles Léopold Mayer, Lausanne, Paris, 2018)
14. OFEN, *Statistique globale suisse de l'énergie 2016* (Office fédéral de l'énergie, Berne, 2017)
15. OFEV, *Indicateurs de l'évolution des émissions de gaz à effet de serre en Suisse 1990–2015* (Office fédéral de l'environnement, Berne, 2017)
16. OFEV, *Pollution sonore en Suisse. Résultats du monitoring national sonBASE, état en 2015* (Office fédéral de l'environnement, Berne, 2018)
17. OFS, *La pendularité en Suisse 2016* (Office fédéral de la statistique, Neuchâtel, 2018)
18. OFS, ARE, *Comportement de la population en matière de transports: Résultats de microre-censement mobilité et transports 2015* (Office fédéral de la statistique et Office fédéral du développement territorial, Neuchâtel, Berne, 2017)

19. R. Oldenziel, M. Emanuel, A.A. de la Bruhèze, F. Veraart (eds.), *Cycling Cities: The European Experience; Hundred Years of Policy and Practice* (Foundation for the History of Technology, Eindhoven, 2016)
20. C. Reid, *Bike Boom: The Unexpected Resurgence of Cycling* (Island Press, Washington, DC, 2017) https://doi.org/10.5822/978-1-61091-817-6
21. B. Spielmann, *'Im Übrigen ging man zu Fuss': Alltagsmobilität in der Schweiz von 1848 bis 1939* (LIBRUM, Basel, 2020)

Part I
Theoretical Framework and Context

Chapter 2
Velomobility

How do we study the practice of cycling? How can we interpret the varying propensity to cycle exhibited by individuals, between territories and over time? Several authors have emphasised the importance of a global approach to studying mobility habits. Such an approach is especially justified in the case of cycling, as the conventional models of transport planning and modal choice analysis—based on minimising costs and travel time—are not sufficient.

I propose to conceptualise velomobility as being made up of three dimensions. In this theoretical framework, (1) the use of the bike (profile of cyclists, characteristics of journeys) is seen as resulting from the meeting of (2) the cycling potential of individuals (access, skills and appropriation related to the bike and other means of transport) and (3) the hosting potential of territories, or their bikeability (spatial structure, infrastructures, norms and rules).

2.1 Analysis of the Practice of Cycling

Commuting by bicycle has traditionally been addressed according to five groups of determinants: the built environment, the natural environment, socio-economic variables, psychological factors, and aspects related to cost, time, effort and safety [33, 36].

The first three determinants have usually been addressed from a macro-perspective, where cycling is analysed on an aggregate level. Its modal share is put into perspective using variables linked to the urban form (density, size, etc.), the presence and quality of infrastructures (e.g. [9, 78] and climate and weather characteristics [65]. Other studies focus on the characteristics of bicycle commuters (such as age, gender and socio-economic status) and the way they differ between countries and cities (e.g. [29].

© The Author(s), under exclusive license to Springer Nature Switzerland AG 2021
P. Rérat, *Cycling to Work*,
SpringerBriefs in Applied Sciences and Technology,
https://doi.org/10.1007/978-3-030-62256-5_2

The last two determinants imply a micro-analytical perspective centred on individuals and their decisions. It may address their revealed (e.g. actual behaviour) or stated (e.g. intentions) preferences as well as the impacts of events promoting cycling. One of the core approaches, for example, centres on psychology-based individualist models of human behaviour—based on the theory of planned behaviour, for example—where the focus is on the individual's attitudes and perceived social norms (e.g. [66]). Another approach is based on utility theory (neo-classical approach) and assumes that an increase in the time, cost and effort of a travel option will result in a decrease in the likelihood of this option being chosen [36].

Several researchers have called for a broadening of the scope of the analysis of bicycle commuting. On the individual level, for example, more qualitative and ethnographic accounts have highlighted the importance of mobility as an embodied experience and as carrying various meanings [73, 89, 94]. On a macro-scale, historical studies have analysed how the practice of cycling depends on material conditions (urban form, infrastructure, amenities) and has very different meanings depending on time, context and social group [13, 37, 46, 69, 78, 80, 89, 96]. The different historical phases mentioned at the beginning of this book each correspond to a specific combination of practices, material conditions and meanings. Thus, in the nineteenth century, the bicycle was reserved for the elites and was a symbol of progress and freedom. It was then democratised in the first half of the twentieth century, before the arrival of the car and the proliferation of motorcycles made it an anachronistic means of transport and caused its decline. It was not until the 1970s, notably due to the emergence of environmental concerns, that cycling began to regain a positive image, at least among certain sections of the population.

Some authors have applied the framework of social practice theory to cycling and analysed it as a combination of materials, competences and meanings [88, 90, 91]. Others have proposed the concept of 'cycling culture' [13] to designate this tangle of dimensions. There are so-called *mature* cycling cultures, such as the Netherlands and Denmark. In these countries, travelling by bicycle is ubiquitous and commonplace, and the bicycle is considered to be an efficient and rapid means of transport [23], even though some practices, such as using a cargo-bike, have connotations in terms of gender and class [7]. However, in most Western countries with a cycling culture that is emerging or being redefined, people who cycle can be regarded as a minority or an out-group in a car-centric world [75]. This is even more the case in developing countries, where the image of the bike remains negative and is usually seen as a low-status means of transportation [95].

We find similar considerations in John Urry who, when studying the car, proposes the concept of the automobile system or automobility. With this term, he highlights the fact that the automobile is much more than a simple vehicle. More generally, it is a socio-technical assemblage, which brings together cars, industries, infrastructure, amenities, rules, images, representations, practices, policies, and so on [19, 87, 93]. By analogy, some authors have put forward the terms 'velomobility' [4, 53, 89] or 'bicycle system' [38] to designate the set of socio-technical elements, which make up and influence the practice of cycling.

Velomobility is seen as an incomplete system that lacks dedicated infrastructures and social legitimacy in a context dominated by automobility [4, 27, 37, 53]. Indeed, automobility and velomobility 'compete for people's time, for road space, for resources, and in discourse' [97], 121), and automobility still has an 'enormous competitive advantage in recruiting practitioners and sustaining performances" in many countries' (ibid. 124).

Frédéric Héran, in his book tracing the history of the bicycle, believes that analysis of the bicycle system must be supplemented by an approach, which he describes as *omnimodal*. He emphasises the need to reposition the evolution of bicycle use within the more general development of the various modes of travel and not to consider it in isolation [37], 14). The total number of journeys per person per day averages at between three and four, and this indicator appears to be stable over time. The modes of transport are thus in competition within a journey market which is not expandable.

These approaches have prompted us to adopt a systemic perspective on cycling. Inspired by the work of Vincent Kaufmann and its conceptualisation of motility [48–50], we believe that velomobility is made up of three main dimensions. The intensity of bicycle use is thus explained by the confluence of the other two components of the reading grid Fig. 2.1: the cycling potential of individuals (individual dimension) and the cycling hosting potential of the territory (contextual dimension). This approach shares principles with the others mentioned above, such as the need for a holistic understanding of cycling. It presents the advantage of enabling the identification of the various mechanisms of the (non-) adoption of cycling both at an individual and a contextual level.

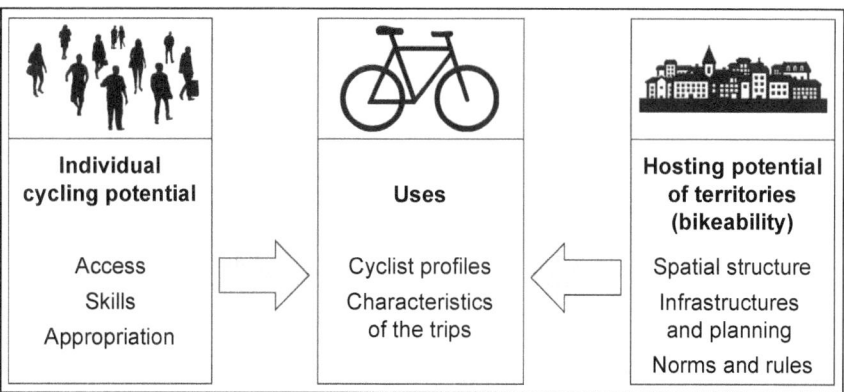

Fig. 2.1 The three dimensions of the system of velomobility (own elaboration; images taken from pixabay.com)

2.2 Uses

The first dimension of the analytical framework consists of the various uses of bicycles that can be identified, in terms of the characteristics of the users and their journeys. This is the traditional approach to transport studies and covers elements of an essentially factual nature.

The user profile includes both socio-demographic (gender, age, type of household, national origin, etc.) and socio-economic (level of education, employment status, social class, etc.) variables.

Journeys undertaken by bicycle can be described on the basis of four distributions. Temporal distribution relates to the frequency with which bicycles are used on a weekly or annual scale. Spatial distribution corresponds to the origins and destinations of the journeys made and the distance travelled. Third, we have causal distribution, i.e. the reasons for which bicycles are used. A distinction is generally made between recreational cycling (journeys made for the sole reason of leisure or sports) and utility cycling (journeys made for a practical reason). In the latter case, cycling represents a means of transport that makes it possible to commute, go shopping, etc. The modal distribution of journeys allows us, ultimately, to bring into focus all journeys that are made using a form of active mobility (cycling, walking, etc.), public transport (bus, tram, train, etc.) or individual motorised transport (car, motorcycle, etc.).

The characteristics of the users and their journeys raise important issues and refer notably to two debates. The first one is how cycling may be inclusive and how the profiles of cyclists may differ in terms of gender, class, age and national origin [2, 29, 39]. The second is the impact of the bike in terms of modal shift and its potential to reduce car journeys and ownership [10, 21, 28, 56, 98].

2.3 The Cycling Potential of Individuals

The mobility potential of an individual—or motility—is defined as the set of characteristics specific to an actor, which allow them to be mobile [48–50]. Individuals are characterised by their aptitude for movement in a given physical, economic and social context. This implies that mobility is thought of not only in terms of the journeys themselves but also in terms of experiences, representations and the capacity and aptitude to be mobile. Motility is built around three dimensions that can be defined in the case of cycling—access (to be able to), skills (to know how to) and appropriation (to want to).

2.3.1 Access

Access ('to be able to') refers to all the mobility options available to a person at a given time and place These facilities can be considered to be a personal portfolio of access rights [24], representing a more or less diversified selection. They include the possession of vehicles (car, motorcycle, bicycle, etc.) and subscriptions to mobility services (public transport, car-sharing system, bike share network, etc.)

In the case of cycling, it amounts to an individual having access to a functional bicycle that corresponds to their needs. According to the Mobility and Transport Microcensus, 76% of residents in Switzerland have a bicycle available to them either permanently or on demand [68]. For some people, however, the financial aspect is a barrier, despite the fact that second-hand bicycle markets organised in many cities make it possible to buy a bicycle at a low price. For others, the risk of theft or damage represents a barrier to cycling. In 2013, almost 40,000 bicycle thefts were reported to the police in Switzerland, a figure that may be doubled if we take into account the number of unreported thefts. The crime-solving rate is 1.3%, or fewer than 500 cases across the country [3]. This shows the crucial importance of adequate parking conditions in order to guarantee long-term access to a bicycle in working condition (see Sect. 2.4.2).

From its invention in the nineteenth century, the bicycle barely changed, until recent years, when technical innovations have resulted in significant changes, leading to a reconfiguration of practices and an opening up of new perspectives. Several trends widen the possible uses of the bicycle, such as the proliferation of folding bicycles (which can easily be taken onto public transport), cargo bikes (which allow the transportation of children or objects), bike-sharing schemes (which can supplement public transport networks and overcome home parking problems, for example), and adaptative bikes (bicycles adapted to certain disabilities).

The main development is the electrically assisted bicycle,[1] which has been very successful in Switzerland. With 133,000 units sold in 2019, it represents almost 40% of new bicycles brought to the market.[2] A section on electrically assisted bicycles was also introduced in the 2015 Mobility and Transport Microcensus. According to this source, 7% of households have at least one electrically assisted bicycle [68].[3] Given its importance, the electrically assisted bicycle deserves particular attention, in both the theoretical and empirical parts.

The electric assistance may contribute to redefining the characteristics of the practice of cycling as a whole: 'the speed of the e-bikes reduces the time required

---

[1] The terms "electric bike" or "e-bike" are more common but are imprecise, as the electric assistance is only delivered when the user pedals and provides an effort themselves.

[2] See www.velosuisse.ch.

[3] Two types of electrically assisted bicycles are distinguished in Swiss legislation. First, electrically assisted bicycles, where the assistance stops at 25 km/h. These are considered light mopeds, a category akin to traditional bicycles. Second, electrically assisted bicycles, where the assistance goes up to 45 km/h. Included in the category of mopeds, they require a licence plate and a driving licence (for two-wheelers with a light engine), as well as the wearing of a helmet. In Switzerland, more than 80% of electrically assisted bicycles sold are limited to 25 km/h.

to travel a given distance or increases the range of travel for a given amount of time relative to conventional bikes. E-bikes also accelerate faster than conventional bikes, and accelerating to and maintaining top speeds require less physical exertion' [74], 39). Because of its 'combination of leg and battery power' [4], 64), the e-bike could have an 'intermediator role' [98] or represent a 'transitional step' [74] between conventional bikes and cars. Moreover, the electric assistance reduces the barrier of topography, effectively 'flattening' topography and making it easier for cyclists to travel further [58, 21]. The electrically assisted bike is thus frequently used for longer distances [10, 45].

The e-bike may enable more people to cycle, including some who could or would not otherwise make the same journey by conventional bike [10, 21, 74]. Several researchers have addressed the profile of e-bike users, although it should be noted that the differences they observed may be due to the spatial context and the period of analysis. Men are more numerous than women in several studies [58, 98], although the opposite is also observed by some researchers [35]. Retired people represent the majority of e-bike users in some studies [98], as the electric assistance makes it possible to go on cycling despite physical decline due to age [57], but in other studies it is people in the second stage of their working life (40–65) who are overrepresented [58]. In terms of social class, e-bike users tend to have an above-average level of education and income (ibid.). This may be due to the price of an e-bike, as well as to the fact that those with a higher level of education are often observed among the early adopters of an innovation [47].

2.3.2 Skills

The practice of a means of transport requires not only access but also skill on the part of the user ('to know how to'). Skills are often underestimated in mobility research, and the research on cycling is generally no exception. However, mobility is learned and experienced, and necessitates a variety of skills and knowledge that correspond to the skills required for the appropriation of a mode of transport.

Of course, knowing how to cycle is the basic skill, and this is widespread among the population. While figures aren't available for Switzerland, almost all French people (97.9%) have learned to ride a bicycle [67], and it is very likely that the Swiss figures are as high. While learning generally occurs in a family setting, lessons at school or in an extra-curricular context allow children to fully master the action of cycling. However, while learning to cycle seems still to be a rite of passage, the actual practice during childhood and adolescence becomes less common [85], which poses a challenge for the recruitment of new cycling practitioners. Other training courses are intended for adults who did not learn to ride a bicycle in their youth (for an analysis of the courses offered in Switzerland and their impact see Mundler and Rérat 2018a, 2018b).

Travelling by bicycle also requires other essential skills. These are far from being common to the whole population, especially when the traffic conditions involve riding alongside motor vehicles. Thus, despite appearances, travelling by bicycle is far from a trivial act and involves skills which fall within five main areas of competence [24]: (1) having the necessary physical abilities, (2) gaining experience in specific traffic situations (involving motorised traffic), (3) having good knowledge of the territory (finding an appropriate route, avoiding natural obstacles or fragmentation of the urban landscape), (4) being able to estimate the duration of journeys and (5) having practical knowledge (orientation, planning a sequence of activities, carrying out repairs,[4] etc.). This last point refers to the notion of 'convivial tool', which Illich defines in his critique of the industrial system as a tool that enhances the 'independent efficiency' of its users and enables them to 'master' themselves [40]. He uses the mechanical bike as an example of convivial tool.

Some consider these skills to be tactics [11] for adapting to a territory, which is still largely designed for automobiles. They are likely to develop depending on the frequency of use. In the adoption of a social practice, it is possible to distinguish a 'trajectory' of individuals who progressively pass from the status of beginner to that of experienced user or expert [88]. The skills required depend on several factors, such as the perception of danger and character traits, but also cohabitation with motorised traffic and the quality of the infrastructure and amenities. Skills are thus crucial because they determine the individual's level of comfort on a bicycle and influence the choice of whether to use this means of transport.

It is important, however, to note that the skills for utility cycling do not only refer to individual characteristics. They are highly dependent on what the hosting potential of a space requires. In other words, increasing the modal share of cycling does not necessarily imply an increase in the level of skills of cyclists, but does require the provision of infrastructures that make cycling safe and appealing for a large population, across age and gender (this point will be discussed in Sect. 2.4.2).

2.3.3 Appropriation

The third motility variable is appropriation ('to want to'), which is defined as the way in which individuals perceive and select the mobility options available to them according to their needs, aspirations, strategies, values and habits. The image that individuals have of modes of transport is key in this regards. It is through this dimension that a latent ability to move is transformed into an effective journey. It is not enough for a mode of transport to be available; it must be used and appropriated [48]. Individuals' appropriation of a mode of transport depends on their perception of that mode and of its particularities. Drawing on Cresswell, perceptions and images of the bicycle can be interpreted as a confluence of three fundamental dimensions of mobility—movement, meaning and experience—in a context of power (here in regards to the dominant system of automobility) [14, 15].

---

[4]We sometimes speak of *velonomy* to mean autonomy in the maintenance and repair of a bicycle.

The first dimension of mobility is physical movement, the most readily visible aspect of mobility. Physical movement constitutes the simple fact of going from A to B and refers to elements that can be easily measured (distance, speed, frequency, reason for travel, etc.). Movement is traditionally the focus of transport planning, which conceives mobility as either meaningless or as the practical outcome of 'rational' decision-makers who optimise variables such as time, cost and distance [89]. Without downplaying the deterring role of distance and time in cycling, this theoretical stance has received two main critiques. First, conventional models of modal choice are not sufficient for cycling, particularly given the importance of effort and weather, for example [36, 71]. Second, it is necessary to explore the content of the line between A and B, to go beyond mobility as a rationalised and instrumental practice [89]. The other two dimensions of mobility, meaning and experience, are less tangible and less easily measurable and relate to the content of the line between A and B.

The second dimension refers to the fact that mobility is also loaded with meanings—from an individual and social point of view—that may be found in representations, discourses and narratives about the fact of moving. The literature on automobility has highlighted how the success of the car is also related to the images and imaginaries associated with it [93], as mentioned above, cycling too conveys a wide range of meanings, which vary depending on the place and time [13]. In countries where cycling is not a commonplace and fully normalised mode of transportation, people who cycle can be regarded as a minority or an out-group in a car-centric world [75]. From a more positive point of view, some cyclists feel that they are 'embodying citizenship' by enacting public policy or civic engagement and by reclaiming an alternative to the dominant system of automobility [6, 59]. Thus cycling can sometimes be considered to be a politicised practice (although not always consciously), in the sense that it represents a message or desideratum [26].

The third dimension refers to the experience of mobility, to the way it is lived, felt and embodied by individuals in various circumstances. Mobility is physically implemented in everyday life, created through embodied and sensory engagement with the urban environment and woven into lives in contextually specific and personalised ways [95]. An increasing number of qualitative accounts address the embodied experience of cycling, showing the extent to which cycling is an everyday practice mediated through the senses [73, 94], and demonstrating that the sensory response is 'clearly a factor in motivation for choosing the bicycle as a mode of transportation' [44], 653). Day, for example, has addressed the sense of movement and flow of bicycle couriers, their interactions with the built and natural environment and the way they internalise and play with the rhythms of the street [17]. Some scholars have stressed that cycling enables social interaction and provides opportunities to meet others more than other means of transportation [91]. However, as will be discussed later, cyclists often feel themselves to be vulnerable road users in a challenging sensory environment within a car world [56].

The three components of mobility—movement, meaning and experience—can be evaluated or perceived in either a positive and negative way. They thus constitute motivations for, or barriers to, the appropriation of a specific means of transport [82].

This perspective makes it possible to open the mobility 'black box', and go beyond a utilitarian and rational approach based on cost and distance, to tackle the more intangible aspects behind modal choices, such as sensory and physical aspects [89].

2.4 The Cycling Hosting Potential of Territories

A context offers a specific field of possibilities, and the hosting potential of a space—understood as the physical, built, social and political environment—refers to how receptive, suitable or welcoming it is for certain practices [48]. The territory and its hosting potential influence access, skills and appropriation with regard to the different means of transport, which both compete and complement each other.

The environment's receptiveness relates to the notion of affordance [31], which derives from the verb to afford, meaning both to provide and to be able to do something. Affordance is a debated analytical concept in science and technology studies, critics have highlighted the need to define and operationalise affordance, to analyse the underlying mechanisms and to account for the diversity of subjects and circumstances [16]. Affordance is relational in that it links the suitability of a context for a use with the intentions and capabilities of potential users. A territory is characterised by affordances (e.g. infrastructures, social norms, laws), which facilitate certain modal choices, while the absence of some affordances may dissuade individuals from choosing some modes. This refers to the notion of friction [15], which, as a social and cultural phenomenon attached to mobility, draws attention to the ways in which people are slowed down or stopped in their practices. Thus in addition to affordances, a space can also be characterised by friction effects [15], or a certain level of viscosity [22].

Applied to cycling, this concept designates the degree of 'bikeability' of a territory, i.e. its adaptation or suitability for the use of bicycles. I use the term 'bikeability' by analogy with the more widespread term of walkability (attractiveness of a space for walking).[5] It has three main aspects: the spatial context, amenities and infrastructure, and non-material elements, such as rules and norms.

2.4.1 Spatial Context

Terrain and climate are frequently given as reasons for a poor modal share of cycling. Of course, a lack of hills makes cycling easier and reduces the physical effort required (except where strong wind creates virtual hills, as is sometimes the case in the Netherlands); however, topography is not always an adequate explanation for cycling practice: some very flat cities, for example, have a very low modal share of cycling, while

---

[5]The term does not refer to individuals, unlike the British initiative of the same name, which promotes training programmes.

others, which are more hilly, have a greater number of cyclists [37]. Furthermore, in the Swiss context, Geneva and Bern have relatively similar topography (and size), but the share of bicycle use in the latter is more than double that of the former (see Sect. 3.2).

The same caution is required when considering the effect on cycling practice of weather conditions (on a daily basis) and climatic conditions (on a seasonal level). According to Héran [37], 'One would expect rain, snow and ice to be significant barriers to regular cycling. How do we explain, therefore, why cycling is so popular in Northern Europe and much less common in the South?'. Furthermore, significant differences exist between cities with the same climate. In North American cities, cycling rates are low and decrease sharply in winter, while in Northern Europe, winter has a relatively small impact. This difference may be explained by the quality of the infrastructure and by the maintenance thereof (snow clearance, preventive gritting, etc.) [41]. Other research has shown that bad weather has more influence on recreational cycling and discretionary travel than on commuting [65]. Thus while geographical and physical constraints do play a role, they are far from being as determinative as one would intuitively expect [37].

The bicycle is an autogenic mode of transport, which means that it moves forward as a result of power exerted by its user. Therefore, in addition to topography, another type of spatial friction should be taken into account: that of distance (e.g. [51, 71]. A distance of 5 km—or 7.5 km in the Netherlands, thanks to routes of higher quality— is frequently mentioned as the limit beyond which traditional bicycles become less attractive. This distance increases to 10 or 15 km for electrically assisted bicycles.

Distance relates to the shape and size of cities. Their density, diversity and mix of urban functions, the attractiveness of the landscape and the built environment along cycle routes, are all factors that influence the practice of cycling [33, 34, 36, 78].

In terms of urban planning, for the last 20 years, political debates in both Switzerland and elsewhere have focused on regulating urban sprawl [81]. The criticisms levelled at this lower-density urban form include dependence on cars and the environmental impacts associated with the use thereof. The compact city model, or inward urbanisation as it is called in Switzerland, is promoted in the new version of the Federal Spatial Planning Act. This model of urbanisation involves the coordination of the public transport infrastructures and a greater density, in the aim of ensuring that public transport, cycling and walking are attractive and efficient.

2.4.2   Infrastructure and Amenities

The extent to which different modes of transport are used is the consequence of power relationships in space (which result from how space and budget are allocated to the different modes) and representations in the field of transport planning [53]. Although active mobility is an essential feature of the contemporary discourse on cities, this has not always been the case.

At the end of the Second World War, transport planning echoed the trend of modern urban planning, which notably promoted a separation of functions (living, working, recreation and traffic). This led to the marginalisation of cycling in planning, with transport infrastructure—mainly road infrastructure—designed to transport people and goods quickly and efficiently. Road traffic was streamlined, the city was sectored and crossed by independent networks, and flows were separated according to their speed [70]. The street—which served as a support for social life, and functioned as a meeting place, not only for people but also for the different modes of transport—gave way to the road, to modern arteries along which car traffic must be able to circulate without interruption. According to Le Corbusier, 'The street is no more. The street has become the new city road, arterial road, the speed-way' [55]. A few years later, the French president Georges Pompidou summed up the spirit of the time with this sentence: 'We must adapt the city to the car'. Switzerland has not escaped this trend, even if it is distinguished by a significant weighting in favour of railways and urban public transport [42], whose high quality has made them serious competitors for active mobility, particularly in built-up areas.

The cities and countries that have experienced the most significant upturns in the practice of cycling have not done so by chance. The construction of infrastructure has played a decisive role in increasing the bikeability of the territory [1, 9, 12, 33, 77].[6] Some authors, such as Héran [37], have also demonstrated the significant, even overriding, impact of measures to moderate the speed and volume of car traffic and ensure better cohabitation with cyclists.

Cycle routes have even more impact if they are designed in such a way as to create a dense, coherent and fast network, and if they are physically separated from the traffic. In addition to traditional bicycle lanes and tracks,[7] the number of projects to build express cycle routes—or bicycle highways—as well as green waves—coordination of traffic lights for cyclists—is on the rise. These are intended to encourage the practice of cycling over greater distances by removing or reducing barriers (e.g. building subways or bridges, increasing the continuity of routes and/or giving priority to cyclists, etc.).

In addition to linear infrastructures, certain strategically placed developments have a positive impact on the practice of cycling. Crossroads, for example, are potentially dangerous places; thus in order to help ensure the fluidity and safety of cycle routes, there are a variety of measures, such as the green wave (synchronising traffic lights for bicycles) and ASLs (advanced stop lines enabling cyclists to stop in front of cars at traffic lights).[8]

Another critical aspect is bicycle parking [76], which must be sheltered and secure in order to enable permanent access to a working bicycle. Where parking is well

---

[6]The impact of such measures may, however, be limited according to other constraints or shortcomings in the cycling culture [60, 62, 80].

[7]The former are indicated by road markings; the latter are specific routes that exclude road traffic.

[8]An ASL is a buffer zone between the traffic light and the line at which motor vehicles stop, which allows cyclists to be visible, gives them more time to pull out and reduces the amount of exhaust fumes they breathe in.

placed, it makes the use of bicycles attractive and efficient. This should be provided both in residential locations (bicycles should be easily accessible, as the place of residence is the starting point for many journeys) and at regularly frequented facilities (companies, commercial areas, training institutions, etc.). The same is true for railway stations, which are seeing the development of bike stations, in addition to parking and various maintenance and repair services.[9] Such amenities can also promote the combination of cycling and public transport.[10]

Several organisations have set out general principles for urban cycling. In the Netherlands, five basic requirements have been defined: routes must be direct, comfortable, attractive, safe and consistent. These criteria are broken down according to the type of route, territorial context and traffic volume, in a detailed manual intended for urban and amenity planners [32]. The standard of requirement is very high because the objective is to make cycling accessible and attractive to as many people as possible. According to this guide, a cycling route that is not safe for an 8-year-old child is not a cycling route! This list was reworked by the *Artgineering* firm, which considers that cycling infrastructure must not only be designed from the point of view of its efficiency in getting you from point A to point B but also as one of the elements that contribute to the quality of the space between A and B, and as an opportunity to transform circulation spaces into real public spaces [5].

Improving the quality of cycling infrastructure does not only have an impact on the number and proportion of journeys made by bicycle, but also diversifies the demographic of cyclists in terms of gender,[11] age group and also skill level and motivation. This was notably illustrated by Geller, an urban planner in Portland. He distinguishes between four types of inhabitant in his city, according to their appropriation of cycling and to how vulnerable they feel on the road: the 'strong and fearless' (who feel at ease without specific facilities; less than 1% of the population[12]), the 'enthused and confident' (who require certain amenities; 6%), the 'interested but concerned' (who would be prepared to cycle provided that the infrastructure is sufficiently developed to ensure their safety; 60%) and the 'no way no how' (who can't or don't want to ride a bike; 33%) [20].

On the basis of this typology, it is crucial to provide framework conditions that do not only address the strong and fearless. The priority issue in promoting cycling is to maintain the practice of the two first categories over the life course and to convince an increasing proportion of the third category to make the leap by providing the necessary framework conditions. As stated above, affordance is relational in that it links the suitability of a context for a use with the intentions and capabilities of potential users.

---

[9]More than 50 bike stations were operational in railway stations in Switzerland in 2020 (www.vel ostation.ch).

[10]The combination of cycling and public transport is very common in the Netherlands: among rail users, 40% cycle to the station and 10% use a bicycle after getting off the train [37], 180).

[11]The proportion of female cyclists is considered to be an indicator of the quality of the infrastructure. In countries and cities where the modal share of bicycles is high, the proportion of women is also high. Conversely, when cycling is not widespread, the share of women is low [29].

[12]The percentages put forward in the case of Portland vary according to context.

If this view is now largely shared among cycling scholars and advocates, this has not always been the case, as shown by the doctrines of Vehicular Cycling (elaborated by John Forrester in the USA) and of Cyclecraft or the Right to Ride (put forward in the UK by John Franklin). These doctrines were highly influential in the 1970 and 1980s and still are, although now to a much lesser extent. They have in common the belief that cyclists should be treated as the drivers of vehicles ('bicycle drivers'), do not need segregated infrastructures and should acquire specific skills, such as increasing their cadence as a means of acceleration out of trouble. According to the Cyclecraft textbook, 'a good cadence to aim for is about 80 rpm [number of turns in one minute], while sprint speed of 32 kph will enable you to tackle most traffic situations with ease' [25]. These principles have been highly criticised, as they limit everyday cycling to just a handful of people (usually young and male), and have failed to increase the modal share of cycling and the sense of safety of riders [12, 80, 96].

2.4.3 Rules and Norms

The affordances linked to the territory's hosting potential are also intangible and relate to legal and social rules and norms. The variety of cycling practices springs from profoundly differing experiences, personal and collective, that are shaped by national history, class, gender and ethnicity [13], 3). Differences in terms of barriers are hence found between spaces and points in history, and also between social groups.

The frictions faced by cycling are therefore also symbolic, first of all in contexts where the system of automobility has been dominant for decades. As the use of the car increased in the first half of the twentieth century, rules were defined. A first series of measures restricted speed and use, and then the highway code was developed, in order to ensure the coexistence of users [52] but also to discipline non-road traffic. Relegated to the pavements and dedicated passages, pedestrians had to give way to the car. The first chinks in this system appeared with the establishment of the first pedestrianised areas in the 1960s. Thirty years later, cars lost priority at pedestrian crossings in Switzerland, where 30 km/h zones and meeting zones[13] have been put in place.

Other rules have been, or are in the process of being, adapted to take into account the specificity of cyclists, such as contra-flow cycling,[14] turn-right-at-red policies[15] and dead ends with exemptions.[16] These measures, which are easy and inexpensive

---

[13] 30 km/h zones are usually in residential areas, and priority is to the right. In a meeting zone, the speed is lowered to 20 km/h and pedestrians have right of way.

[14] Streets with one-way traffic for motor vehicles and two-way traffic for bicycles.

[15] Transformation of red lights for cyclists by giving way to them at certain crossroads. This measure, which is in force in several countries, has been tested conclusively in Basel and will be incorporated into the Swiss Federal Road Traffic Act.

[16] These indicate permeability for bicycle traffic and pave the way for practical and sometimes unknown routes. For example, in the Swiss city of Lausanne, a quarter of the 172 listed dead-ends are permeable to bicycles.

to implement, have a triple interest for cyclists: safety (better visibility and ability to choose the least dangerous routes), diversity (of possible destinations and routes) and speed (more direct journeys, requiring less effort). In this regards, it should be noted that for a cyclist travelling at 20 km/h, restarting after stopping represents an extension of the journey by about 80 m, while climbing one metre corresponds to a detour of about 50 m [37], 49).

Rules are not only written explicitly into official laws and ordinances but also implicitly into the fabric of society. These are the standards and norms in which individuals have internalised, whether consciously or not, and which influence their perceptions and behaviours. In the case of mobility, the images and values associated with different modes of transport make them more or less attractive, desirable or legitimate. Social norms thus influence the appropriation of different means of transport by individuals. The cultural meanings associated with the car (as a symbol of freedom or of social status, etc.) participated not only in its development but also in the decline or depreciation of other modes [93].

Lee believes that the car has informally privatised public space such that other users no longer feel legitimate, and the street has become a dangerous terrain [56]. According to Prati and her colleagues, where cycling is developed, it is accepted, where it is rare, it is less tolerated and subject to negative attitudes, as the minority practice can be seen as a critique of the majority position held by the automotive system [75]. Handy et al. [33] observe the impact of the social environment (friends, family and social norms): if cycling is perceived as a normal and legitimate way to travel, residents are more inclined to use it themselves, thereby helping to increase its popularity.

With the idea of social norms, we come back to the notion of cycling culture, as presented in the introduction to the chapter. The positioning of the bicycle is due to a mixture of political will, exogenous constraints and favourable sociological circumstances [79]. Héran [37] warns against any 'cultural determinism' in the interpretation of the differences between Northern countries and Latin countries, or between the linguistic regions of Belgium or Switzerland. As we will see in the next chapter, there are great disparities between cities and regions located within the same European country.

2.5 Cycling Promotion Campaigns

Daily mobility practices are likely to change according to key events in the residential (relocation), social and familial (having children, etc.) and professional (new job, increase in income, etc.) trajectories. As shown by the biographical approach [54, 86], the adoption and continuation of mobility practices are often not linear. Cycling may be maintained throughout the life course, but may also be abandoned or resumed depending on circumstances [43].[17]

Changes in behaviour may also be triggered by promotion campaigns that take a great variety of forms. Some diffuse information to a general public or to a specific target group (e.g. within a company); some others, which are of interest here, encourage participants to try cycling for a certain duration (from one day to several weeks), for certain motives (utility, leisure, or sport) or purposes (environment, health, etc.) (e.g. [18, 56, 72].

Existing research usually addresses event-based behaviour changes through interviews and surveys before, during and after the experience, sometimes in comparison to a control group [99]. Core questions usually centre around who participates in these actions, why they participate, and what impact participation has on mobility behaviour. While Piatkowski et al. [72] find no evidence to suggest that the *bike to work* day in Denver, USA, caused significant change with regards to cycling frequency, other research studying longer events that are more focused on a particular kind of user (e.g. car owners to whom an e-bike is lent) or that imply various incentives (e.g. prizes) shows positive impacts, albeit to varying extents [8, 61, 77, 84, 99].

Some studies go into more detail regarding the practice of cycling, analysing self-reported skills and confidence [8, 92], satisfaction with travel [18], or health indicators [83]. The impact of cycling events may go beyond bike use, to have an influence on self-confidence, sense of belonging, and empowerment, as shown in the case of cycling lessons attended by women migrants [63, 64].

Another approach centre on psychology-based individualist models such as Ajzen's theory of planned behaviour looks at how behaviours are formed and linked with intentions (e.g. [66].[18] Types of (potential) cyclists are also distinguished on the basis of Prochaska's transtheoretical model and on the various stages of change: precontemplation (unaware of problems, no intention to change), contemplation (aware of problems, thinking about change), prepared for action (intention to change in the next 6 months), action (action being taken), and maintenance (has maintained action for 6 months or more) (e.g. [30]. In this perspective, modal shift is a slow process and each stage requires specific strategies, including those that help individuals to maintain new mobility practices.

As discussed above, this book takes a different view of the concept of velomobility. Although the main objective is to provide a global analysis of the practice of utility cycling, a secondary focus will address the impacts of the *bike to work* campaign. These impacts may refer to any of the dimensions of our analytical grid: uses, the cycling potential of individuals or the cycling hosting potential of territories.

---

[17] In their study of seniors, Jones et al. [43] identify, for example, three cycling trajectories: resilient (continuation of the practice), restorative (resumption after a short or long break during their working life) and diminutive (progressive abandonment).

[18] Ajzen's theory states that three elements determine intentions and behaviours: attitude towards the behaviour (e.g. positive or negative evaluation), subjective norms (linked to social pressure), and perceived behavioural control (difficulty and feasibility of alternative behaviours) (Al Chalabi, 2013).

* * *

In general, the rise of the bicycle is the result of the historical and socio-cultural context, and of political choices. The theoretical framework that we have discussed in this chapter has been adapted to analyse bicycle commuting in Switzerland, and has structured the survey which is the subject of this work. The analysis successively addresses the three main dimensions of travel, the cycling potential of individuals (and more particularly their transportation equipment, their skills, and the motivations and barriers conditioning their use of bicycles) and the cycling hosting potential of territories (spatial context, bikeability of commuting journeys, and the intervention of public authorities). Before this, the next chapter puts forward some statistics in order to characterise the practice of cycling in Switzerland and to put it into perspective on an international scale.

References

1. R. Aldred, B. Elliott, J. Woodcock, A. Goodman, Cycling provision separated from motor traffic: a systematic review exploring whether stated preferences vary by gender and age. Trans. Rev. **37**(1), 29–55 (2017). https://doi.org/10.1080/01441647.2016.1200156
2. R. Aldred, J. Woodcock, A. Goodman, Does more cycling mean more diversity in cycling? Trans. Rev. **36**(1), 28–44 (2016). https://doi.org/10.1080/01441647.2015.1014451
3. D. Balmer, Nous avons remonté la piste de nos vélos, tous volés. *Le Matin Dimanche*, 8–9 (2014, September 21)
4. F. Behrendt, Why cycling matters for electric mobility: towards diverse, active and sustainable e-mobilities. Mobilities **13**(1), 64–80 (2018). https://doi.org/10.1080/17450101.2017.1335463
5. S. Bendiks, A. Degros, *Cycle Infrastructure* (nai010 publishers, Rotterdam, 2013)
6. J. Bonham, B. Koth, Universities and the cycling culture. Trans. Res. Part D Trans. Env. **15**(2), 94–102 (2010). https://doi.org/10.1016/j.trd.2009.09.006
7. W.R. Boterman, Carrying class and gender: cargo bikes as symbolic markers of egalitarian gender roles of urban middle classes in Dutch inner cities. Soc. Cult. Geograp. **1–20** (2018). https://doi.org/10.1080/14649365.2018.1489975
8. H.R. Bowles, C. Rissel, A. Bauman, Mass community cycling events: Who participates and is their behaviour influenced by participation? Int. J. Beh. Nut. Phy. Act. **3**(1), 39 (2006). https://doi.org/10.1186/1479-5868-3-39
9. R. Buehler, J. Dill, Bikeway networks: a review of effects on cycling. Trans. Rev. **36**(1), 9–27 (2016). https://doi.org/10.1080/01441647.2015.1069908
10. S. Cairns, F. Behrendt, D. Raffo, C. Beaumont, C. Kiefer, Electrically-assisted bikes: potential impacts on travel behaviour. Trans. Res. Part A Policy Prac. **103**, 327–342 (2017). https://doi.org/10.1016/j.tra.2017.03.007
11. M. de Certeau, *The Practice of Everyday Life* (University of California Press, Oakland, 2013)
12. M. Colville-Andersen, *Copenhagenize: The Definitive Guide to Global Bicycle Urbanism* (Island Press, Washington DC, 2018) https://doi.org/10.5822/978-1-61091-939-5
13. P. Cox (ed.), *Cycling Cultures* (University of Chester Press, Chester, 2015)
14. T. Cresswell, *On the Move: Mobility in the Modern Western World* (1Routledge, New York, 2006)
15. T. Cresswell, Towards a politics of mobility. Env. Plan. D Soc. Space **28**(1), 17–31 (2010). https://doi.org/10.1068/d11407
16. J.L. Davis, J.B. Chouinard, Theorizing affordances: from request to refuse. Bull. Sci. Technol. Soc. **36**(4), 241–248 (2016). https://doi.org/10.1177/0270467617714944

17. J. Day, *Cyclogeography: Journeys of a London Bicycle Courier* (Notting Hill Editions, London, 2015)
18. J. de Kruijf, D. Ettema, C.B.M. Kamphuis, M. Dijst, Evaluation of an incentive program to stimulate the shift from car commuting to e-cycling in the Netherlands. J. Trans. Health **10**, 74–83 (2018). https://doi.org/10.1016/j.jth.2018.06.003
19. K. Dennis, J. Urry, *After the Car* (Polity Press, Cambridge, 2009)
20. J. Dill, N. McNeil, Four types of cyclists?: examination of typology for better understanding of bicycling behavior and potential. Trans. Res. Record J. Trans. Res. Board **2387**, 129–138 (2013). https://doi.org/10.3141/2387-15
21. J. Dill, G. Rose, Electric Bikes and transportation policy: insights from early adopters. Trans. Res. Record J. Trans. Res. Board **2314**, 1–6 (2012). https://doi.org/10.3141/2314-01
22. C. Doherty, Agentive motility meets structural viscosity: Australian families relocating in educational markets. Mobilities **10**(2), 249–266 (2015). https://doi.org/10.1080/17450101.2013.853951
23. E. Fishman, Cycling as transport. Trans. Rev. **36**(1), 1–8 (2016). https://doi.org/10.1080/01441647.2015.1114271
24. M. Flamm, *Comprendre le choix modal: les déterminants des pratiques modales et des représentations individuelles des moyens de transport* (Ecole Polytechnique Fédérale de Lausanne, Lausanne, 2004)
25. J. Franklin, *Cyclecraft: The Complete Guide to Safe and Enjoyable Cycling for Adults and Children* (TSO, London, 2007)
26. Z. Furness, Critical mass, urban space and vélomobility. Mobilities **2**(2), 299–319 (2007). https://doi.org/10.1080/17450100701381607
27. Z.M. Furness, *One Less Car: Bicycling and the Politics of Automobility* (Temple University Press, Philadelphia, 2010)
28. A. Fyhri, E. Heinen, N. Fearnley, H.B. Sundfør, A push to cycling—exploring the e-bike's role in overcoming barriers to bicycle use with a survey and an intervention study. Int. J. Sustain. Trans. **11**(9), 681–95 (2017). https://doi.org/10.1080/15568318.2017.1302526
29. J. Garrard, S. Handy, J. Dill, Women and cycling, in *City Cycling*. ed. by J. Pucher, R. Buehler (MIT Press, Cambridge, 2012), pp. 211–234
30. B. Gatersleben, K.M. Appleton, Contemplating cycling to work: attitudes and perceptions in different stages of change. Trans. Res. Part A Policy Prac. **41**(4), 302–312 (2007). https://doi.org/10.1016/j.tra.2006.09.002
31. J.J. Gibson, *The Ecological Approach to Visual Perception* (Taylor & Francis, Hoboken, 2014)
32. R. de Groot (ed.), *Design Manual for Bicycle Traffic*, Revised. (CROW, Ede, 2016)
33. S. Handy, B. van Wee, K. Kroesen, Promoting cycling for transport: research needs and challenges. Trans. Rev. **34**(1), 4–24 (2014). https://doi.org/10.1080/01441647.2013.860204
34. L. Harms, L. Bertolini, M. te Brömmelstroet, Spatial and social variations in cycling patterns in a mature cycling country exploring differences and trends. J. Trans. Health **1**(4), 232–242 (2014). https://doi.org/10.1016/j.jth.2014.09.012
35. S. Haustein, M. Møller, Age and attitude: changes in cycling patterns of different e-bike user segments. Int. J. Sustain. Trans. **10**(9), 836–846 (2016). https://doi.org/10.1080/15568318.2016.1162881
36. E. Heinen, B. van Wee, K. Maat, Commuting by bicycle: an overview of the literature. Trans. Rev. **30**(1), 59–96 (2010). https://doi.org/10.1080/01441640903187001
37. F. Héran, *Le retour de la bicyclette: une histoire des déplacements urbains en Europe, de 1817 à 2050* (La Découverte, Paris, 2014)
38. F. Héran, The Bicycle System. *Mobile Lives Forum* (2018). https://en.forumviesmobiles.org/marks/bicycle-system-12440. Accessed 1 August 2020
39. M.L. Hoffmann, *Bike Lanes Are White Lanes: Bicycle Advocacy and Urban Planning* (University of Nebraska Press, Lincoln, 2016)
40. I. Illich, *Tools for Conviviality* (Calder and Boyars, London, 1973)
41. E. Jaffe, What the U.S. Can Learn From Northern Europe About Winter Cycling. *CityLab* (2016). https://www.citylab.com/commute/2016/01/winter-bike-riding-seasonal-cycling/426960/. Accessed 1 August 2020

42. C. Jemelin, *Transports publics dans les villes: leur retour en force en Suisse* (Presses polytechniques et universitaires romandes, Lausanne, 2008)
43. H. Jones, K. Chatterjee, S. Gray, A biographical approach to studying individual change and continuity in walking and cycling over the life course. J. Trans. Health **1**(3), 182–189 (2014). https://doi.org/10.1016/j.jth.2014.07.004
44. P. Jones, Sensory indiscipline and affect: a study of commuter cycling. Soc. Cult. Geograp. **13**(6), 645–658 (2012). https://doi.org/10.1080/14649365.2012.713505
45. T. Jones, L. Harms, E. Heinen, Motives, perceptions and experiences of electric bicycle owners and implications for health, wellbeing and mobility. J. Transp. Geogr. **53**, 41–49 (2016). https://doi.org/10.1016/j.jtrangeo.2016.04.006
46. P. Jordan, *In the City of Bikes: The Story of the Amsterdam Cyclist* (Harper Perennial, New York, 2013)
47. K.K. Kapoor, Y.K. Dwivedi, M.D. Williams, Rogers' innovation adoption attributes: a systematic review and synthesis of existing research. Inf. Sys. Manag. **31**(1), 74–91 (2014). https://doi.org/10.1080/10580530.2014.854103
48. V. Kaufmann, *Rethinking the City: Urban Dynamics and Motility* (Routledge & EPFL Press, Lausanne, 2011)
49. V. Kaufmann, M.M. Bergman, D. Joye, Motility: mobility as capital. Int. J. Urban Reg. Res. **28**(4), 745–756 (2004)
50. V. Kaufmann, E. Ravalet, E. Dupuit (eds.), *Motilité et mobilité: mode d'emploi* (Éditions Alphil-Presses universitaires suisses, Neuchâtel, 2015)
51. S. Kingham, J. Dickinson, S. Copsey, Travelling to work: will people move out of their cars. Transp. Policy **8**(2), 151–160 (2001). https://doi.org/10.1016/S0967-070X(01)00005-1
52. A. Kletzlen, L'automobile et la loi. Comment est né le code de la route? Recherche—Transports—Sécurité **68**, 89 (2000). https://doi.org/10.1016/S0761-8980(00)90035-9
53. T. Koglin, T. Rye, The marginalisation of bicycling in Modernist urban transport planning. J. Trans. Health **1**(4), 214–222 (2014). https://doi.org/10.1016/j.jth.2014.09.006
54. M. Lanzendorf, Key events and their effect on mobility biographies: the case of childbirth. Int. J. Sustain. Trans. **4**(5), 272–292 (2010). https://doi.org/10.1080/15568310903145188
55. Le Corbusier, *Sur les 4 routes* (Éditions Gallimard, Paris, 1941)
56. D.J. Lee, Embodied bicycle commuters in a car world. Soc. Cult. Geograp. **17**(3), 402–420 (2015). https://doi.org/10.1080/14649365.2015.1077265
57. S.J. Leger, J.L. Dean, S. Edge, J.M. Casello, "If I had a regular bicycle, I wouldn't be out riding anymore": perspectives on the potential of e-bikes to support active living and independent mobility among older adults in Waterloo, Canada. Trans. Res. Part A Policy Prac. **123**, 240–254 (2019). https://doi.org/10.1016/j.tra.2018.10.009
58. J. MacArthur, J. Dill, M. Person, Electric Bikes in North America: Results of an Online Survey. Transportation Research Record: Journal of the Transportation Research Board **2468**(1), 123–130 (2014). https://doi.org/10.3141/2468-14
59. J. McKenna, M. Whatling, Qualitative accounts of urban commuter cycling. Health Edu. **107**(5), 448–462 (2007). https://doi.org/10.1108/09654280710778583
60. N. Morgan, Cycling infrastructure and the development of a bicycle commuting socio-technical system: the case of Johannesburg. Appl. Mobil. **1–18** (2017). https://doi.org/10.1080/23800127.2017.1416829
61. C. Moser, Y. Blumer, S.L. Hille, E-bike trials' potential to promote sustained changes in car owners mobility habits. Environ. Res. Lett. **13**(4), 044025 (2018). https://doi.org/10.1088/1748-9326/aaad73
62. N. Mueller, D. Rojas-Rueda, M. Salmon, D. Martinez, A. Ambros, C. Brand, A. de Nazelle, E. Dons, M. Gaupp-Berghausen, R. Gerike, T. Götschi, F. Iacorossi, L. Int Panis, S. Kahlmeier, E. Raser, M. Nieuwenhuijsen, Health impact assessment of cycling network expansions in European cities. Prev. Med. **109**, 62–70 (2018). https://doi.org/10.1016/j.ypmed.2017.12.011
63. M. Mundler, P. Rérat, Le vélo comme outil d'empowerment. Les impacts des cours de vélo pour adultes sur les pratiques socio-spatiales. Les Cahiers scientifiques du transport **73**, 139–160 (2018)

64. M. Mundler, P. Rérat, "C'est la liberté !" Etude des cours de vélo pour adultes en Suisse. Etudes Urbain. **3**, 1–62 (2018)
65. M. Nankervis, The effect of weather and climate on bicycle commuting. Trans. Res. Part A Policy Prac. **33**(6), 417–431 (1999). https://doi.org/10.1016/S0965-8564(98)00022-6
66. A. Nkurunziza, M. Zuidgeest, M. Brussel, M. Van Maarseveen, Examining the potential for modal change: Motivators and barriers for bicycle commuting in Dar-es-Salaam. Transp. Policy **24**, 249–259 (2012). https://doi.org/10.1016/j.tranpol.2012.09.002
67. Observatoire des mobilités actives, *Les Français et le vélo en 2012. Pratiques et attentes* (Club des villes & territoires cyclables, Paris, 2013)
68. OFS, & ARE, *Comportement de la population en matière de transports: Résultats de microrecensement mobilité et transports 2015* (Office fédéral de la statistique & Office fédéral du développement territorial, Neuchâtel & Berne, 2017)
69. R. Oldenziel, M. Emanuel, A.A. de la Bruhèze, F. Veraart (eds.), *Cycling Cities: The European Experience; Hundred Years of Policy and Practice* (Foundation for the History of Technology, Eindhoven, 2016)
70. T. Paquot, *Terre urbaine: cinq défis pour le devenir urbain de la planète* (La Découverte, Paris, 2016)
71. J. Parkin, T.J. Ryley, T.J. Jones, Barriers to cycling: an exploration of quantitative analyses, in *Cycling and Society.* (Ashgate, Farnham, 2007), pp. 67–82
72. D. Piatkowski, R. Bronson, W. Marshall, K.J. Krizek, Measuring the impacts of bike-to-work day events and identifying barriers to increased commuter cycling. J. Urban Plan. Devel. **141**(4), 04014034 (2015). https://doi.org/10.1061/(ASCE)UP.1943-5444.0000239
73. C. Popan, Beyond utilitarian mobilities: cycling senses and the subversion of the car system. Appl. Mobil. **1–17** (2020). https://doi.org/10.1080/23800127.2020.1775942
74. N. Popovich, E. Gordon, Z. Shao, Y. Xing, Y. Wang, S. Handy, Experiences of electric bicycle users in the Sacramento California Area. Travel Behav. Soc. **1**(2), 37–44 (2014). https://doi.org/10.1016/j.tbs.2013.10.006
75. G. Prati, V. Marín Puchades, L. Pietrantoni, Cyclists as a minority group? Trans. Res. Part F Traff. Psychol. Behav. **47**, 34–41 (2017). https://doi.org/10.1016/j.trf.2017.04.008
76. J. Pucher, R. Buehler, Making cycling irresistible: lessons from The Netherlands, Denmark and Germany. Trans. Rev. **28**(4), 495–528 (2008). https://doi.org/10.1080/01441640701806612
77. J. Pucher, J. Dill, S. Handy, Infrastructure, programs, and policies to increase bicycling: an international review. Prev. Med. **50**, S106–S125 (2010). https://doi.org/10.1016/j.ypmed.2009.07.028
78. J.R. Pucher, R. Buehler (eds.), *City Cycling* (MIT Press, Cambridge, 2012)
79. O. Razemon, *Le pouvoir de la pédale: comment le vélo transforme nos sociétés cabossées* (Rue de l'échiquier, Paris, 2014)
80. C. Reid, *Bike Boom: The Unexpected Resurgence of Cycling* (Island Press, Washington DC, 2017) https://doi.org/10.5822/978-1-61091-817-6
81. P. Rérat, Housing, the compact city and sustainable development: some insights from recent urban trends in Switzerland. Int. J. Hous. Policy **12**(2), 115–136 (2012). https://doi.org/10.1080/14616718.2012.681570
82. P. Rérat, Cycling to work: meanings and experiences of a sustainable practice. Trans. Res. Part A Policy Prac. **123**, 91–104 (2019). https://doi.org/10.1016/j.tra.2018.10.017
83. C. Rissel, G. Watkins, Impact on cycling behavior and weight loss of a national cycling skills program (AustCycle) in Australia 2010–2013. J. Trans. Health **1**(2), 134–140 (2014). https://doi.org/10.1016/j.jth.2014.01.002
84. G. Rose, H. Marfurt, Travel behaviour change impacts of a major ride to work day event. Trans. Res. Part A Policy Prac. **41**(4), 351–364 (2007). https://doi.org/10.1016/j.tra.2006.10.001
85. D. Sauter, K. Wyss, *Etude pilote sur l'utilisation du vélo chez les jeunes dans le canton de Bâle-Ville* (Département des constructions et des transports & OFROU, Bâle & Berne, 2014)
86. J. Scheiner, C. Holz-Rau, Changes in travel mode use after residential relocation: a contribution to mobility biographies. Transportation **40**(2), 431–458 (2013). https://doi.org/10.1007/s11116-012-9417-6

87. M. Sheller, J. Urry, The city and the car. Int. J. Urban Reg. Res. **24**(4), 737–757 (2000). https://doi.org/10.1111/1468-2427.0027
88. E. Shove, M. Pantzar, M. Watson, *The Dynamics of Social Practice: Everyday Life and How It Changes* (SAGE, London, 2012)
89. J. Spinney, Cycling the city: Movement, meaning and method. . Geograp. Comp. **3**(2), 817–835 (2009). https://doi.org/10.1111/j.1749-8198.2008.00211.x
90. F. Spotswood, T. Chatterton, A. Tapp, D. Williams, Analysing cycling as a social practice: An empirical grounding for behaviour change. Trans. Res. Part F Traff. Psychol. Behav. **29**, 22–33 (2015). https://doi.org/10.1016/j.trf.2014.12.001
91. M. te Brömmelstroet, A. Nikolaeva, M. Glaser, M.S. Nicolaisen, C. Chan, Travelling together alone and alone together: Mobility and potential exposure to diversity. Appl. Mob. **2**(1), 1–15 (2017). https://doi.org/10.1080/23800127.2017.1283122
92. B. Telfer, C. Rissel, J. Bindon, T. Bosch, Encouraging cycling through a pilot cycling proficiency training program among adults in central Sydney. J. Sci. Med. Sport **9**(1–2), 151–156 (2006). https://doi.org/10.1016/j.jsams.2005.06.001
93. J. Urry, The 'system' of automobility. Theory, Cult. Soc. **21**(4–5), 25–39 (2004). https://doi.org/10.1177/0263276404046059
94. J. van Duppen, B. Spierings, Retracing trajectories: the embodied experience of cycling, urban sensescapes and the commute between 'neighbourhood' and 'city' in Utrecht, NL. J. Transp. Geogr. **30**, 234–243 (2013). https://doi.org/10.1016/j.jtrangeo.2013.02.006
95. L.A. Vivanco, *Reconsidering the Bicycle: An Anthropological Perspective on a New (Old) Thing* (Routledge, London, 2013)
96. P. Walker, *Bike Nation: How Cycling Can Save the World* (Yellow Jersey Press, London, 2017)
97. M. Watson, Building future systems of velomobility, in *Sustainable Practices: Social Theory and Climate Change*, ed. by E. Shove, N. Spurling (Routledge, London, 2013), pp. 117–131. https://doi.org/https://doi.org/10.4324/9780203071052
98. A. Wolf, S. Seebauer, Technology adoption of electric bicycles: A survey among early adopters. Transp. Res. Part A Policy Pract. **69**, 196–211 (2014). https://doi.org/10.1016/j.tra.2014.08.007
99. L. Yang, S. Sahlqvist, A. McMinn, S. Griffin, D. Ogilvie, Interventions to promote cycling: systematic review. BMJ, **341**(oct18 2), c5293–c5293 (2010), https://doi.org/https://doi.org/10.1136/bmj.c5293

Chapter 3
Cycling in Switzerland

What is the level of cycling in Switzerland? Many observers have demonstrated the renaissance of the bicycle in Western cities, but what about in Swiss urban centres?

To measure the practice of cycling within the country and to get a more precise picture of its evolution, we first address the proportion of the population who claim to cycle. Next, we present the share of total journeys made by bicycle and compare this across territories, genders and age groups.

3.1 The Majority of the Population Cycles at Least Occasionally

In 2015, a survey conducted by the Federal Statistical Office, involving 3,000 people living in Switzerland and aged between 15 and 74, addressed the frequency of use of different modes of transport. More than a third of the sample say that they cycle on a weekly basis or more frequently (Table 3.1), less than a quarter are occasional cyclists, while 4 in 10 never ride a bicycle.

Marked differences appear between the linguistic regions. In German-speaking Switzerland, the practice is much more widespread. The only category that is under-represented is people who never ride a bicycle (less than one in three). This value is almost 30 points lower than that observed for French-speaking Switzerland and Italian-speaking Switzerland.

3.2 Increasingly Urban Usage of the Bicycle

The Mobility and Transport Microcensus takes into account the different modes of transport and measures the modal share of cycling, i.e. the share it holds of total

© The Author(s), under exclusive license to Springer Nature Switzerland AG 2021
P. Rérat, *Cycling to Work*,
SpringerBriefs in Applied Sciences and Technology,
https://doi.org/10.1007/978-3-030-62256-5_3

Table 3.1 Frequency of bicycle use in the resident population (15–74 years old), 2015

	Every day (%)	Almost every day (%)	At least once a week (%)	At least once a month (%)	Less than once a month (%)	Never (%)	Total (%)
German-speaking Swiss	10.2	9.1	23.3	12.7	12.5	32.2	100
French-speaking Swiss	4.0	3.6	12.2	9.3	12.1	58.9	100
Italian-speaking Swiss	3.4	5.4	12.0	10.3	9.2	59.7	100
Total	8.3	7.6	20.0	11.7	12.3	40.2	100

Source omnibus survey, Federal Statistical Office

journeys in Switzerland. A journey is defined by its reason (going to work or to a shop, visiting someone, getting out in the fresh air, etc.). If more than one means of transport is used during this journey, only the main mode—defined according to potential speed criteria—is selected. Thus a cyclist going to the railway station to take the train is considered to be a user of public transport. Despite this limitation, it is an indicator that reflects the practice of cycling relatively well, and which has the advantage of being able to be positioned within an international perspective.[1]

The proportion of journeys in Switzerland made by bicycle decreased from 1994 (8.7%) to 2010 (6.2%), before experiencing a slight rebound in 2015 (6.8%). According to these statistics, Switzerland appears to be an 'average' country compared with other Western countries (Fig. 3.1). It is thus situated above the English-speaking and Latin countries but below most of the countries of Northern Europe, and in particular Denmark (15%) and the Netherlands (28%). Differences are, however, found between linguistic regions: the modal share reaches 8.6% in the German-speaking cantons (7.7% in 2010) but only 2.9% (2.8%) in the French-speaking regions and 2.7% (2.0%) in the Italian-speaking part.

Figures measured at the national level tend to mask substantial disparities at the regional and local levels. Switzerland is no exception, even though the differences between the types of municipalities are recent (Fig. 3.2). Indeed, in 2010, the share of journeys completed by bicycle differs little from one category to another, even if a slightly higher value is observed in large- or medium-sized cities, and a lower level in the outskirts of major cities (the metropolitan suburbs). Between 2010 and 2015,

---

[1] This measure takes into account the services offered by mobility (travelling to a workplace, for example, irrespective of distance or means of transport used). Other indicators—distance, duration and stage, where 'stage' denotes a defined portion of travel via a mode of transport—are more frequently used in the analysis of the Mobility and Transport Microcensus but also have certain limitations. Distance gives a greater weighting to long-range transport (train and car) and stage to walking, e.g. (including from a parking spot or a public transport station to the final destination). According to these indicators, the share of cycling is 2.4% of distances, 5.2% of total journey duration and 5.3% of stages.

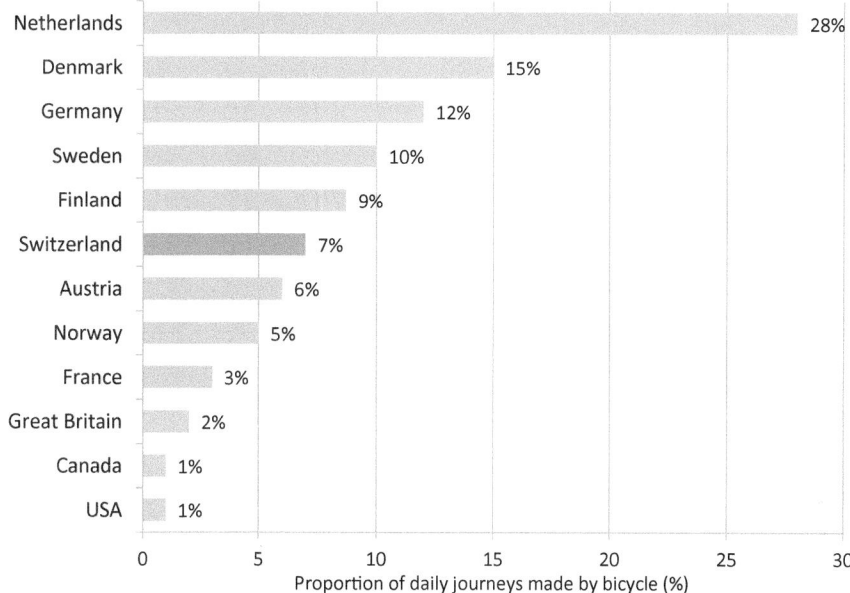

Fig. 3.1 Share of cycling in total journeys in selected countries (2009/2016). *Note* The data come from different sources (city mobility surveys, national statistical offices and European Commission). Methodological differences (data collection, year, spatial breakdown, etc.) limit the comparability of the data

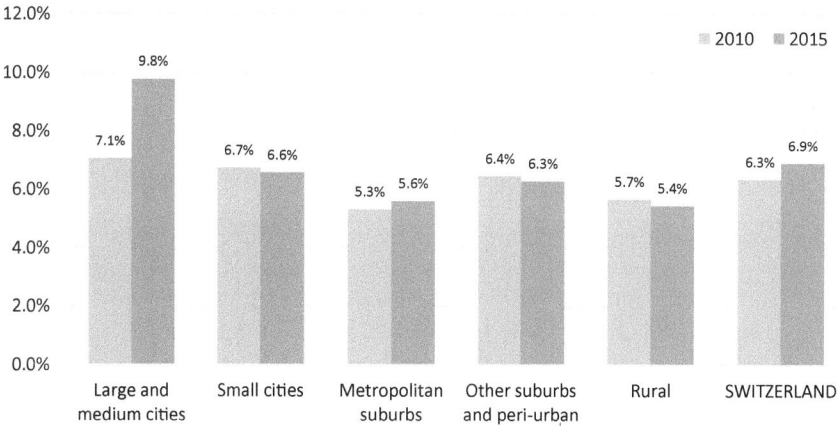

Fig. 3.2 Share of cycling in total journeys by type of municipality, 2010–2015 (*Source* Mobility and Transport Microcensus)

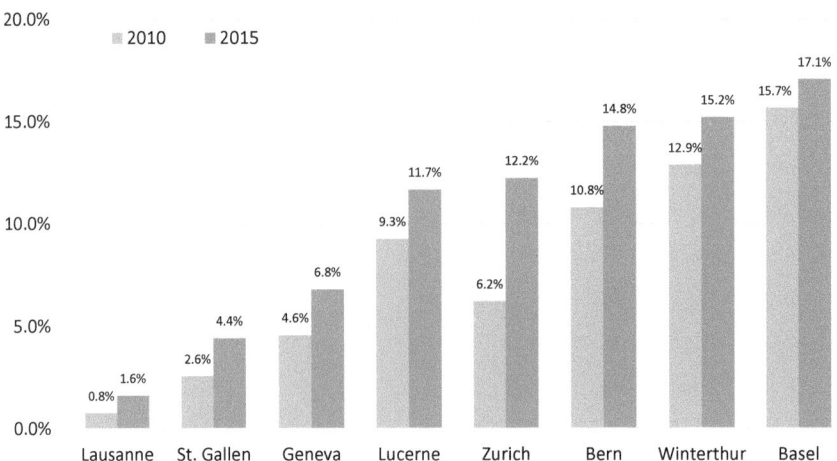

Fig. 3.3 Share of cycling in total journeys in the major Swiss cities, 2010–2015 (*Source* Mobility and Transport Microcensus)

there is a marked evolution of the bicycle in large- or medium-sized urban centres (+2.7 points) and measured growth is observed in the metropolitan suburbs (+0.6). In the other territories, however, the share of cycling decreases by a few tenths of 1% in proportions close to the margin of error.

The practice of cycling has therefore become more urban in recent years, although it is not limited to cities. We can see the consequences of multiple factors, such as a younger population in urban centres [3, 4], the flexibility and efficiency of cycling in the city, congestion of public transport and road infrastructure, public policies limiting the use of the car, etc. In peri-urban and rural areas, as well as in small towns and regional centres, the practice of cycling is not progressing and the car still reigns, given the infrastructure, the distances to be covered, and the relative lack of restrictions on use.

Any conclusions drawn regarding a more urban practice of cycling must, however, be advanced with caution. On this scale also, the disparities are significant. Taking into consideration the largest cities in the country (Fig. 3.3), a ratio of 1 to 10 is observed between Lausanne (1.6%) and Basel (17.1%). Two other cities hover around the 15% mark: Winterthur and Bern. As for Zurich, the 2010–2015 period sees the practice of cycling double. The German-speaking city of St. Gallen, which appears in the rankings between the two French-speaking cities of Lausanne and Geneva, shows that, while disparities exist between linguistic regions, they should be put into perspective according to local characteristics, particularly in terms of transport.[2]

---

[2]The differences between cities are indeed far from resulting only from their topography or linguistic region. They reflect their urban form and density, the distribution of economic activities and housing, the attractiveness of other forms of mobility but also, and perhaps above all, the place of cycling in political priorities and the quality of the amenities, infrastructure and services made available.

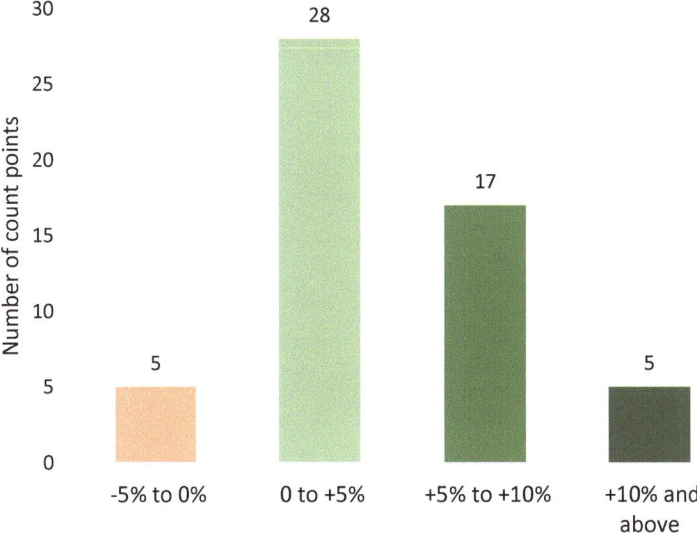

Fig. 3.4 Distribution of bicycle count points in major Swiss cities according to average annual traffic development, 2011–2017 (*Source* Baehler et al. [1])

The count points used by different municipalities are an additional source of information. We have undertaken a census of these approaches in Switzerland [1]. While these are not yet very numerous and are based on variable methods, it is possible to compare the evolution of bicycle traffic in the country's eight largest cities by calculating an average annual growth rate for 55 count points between 2011 and 2017 (Fig. 3.4). In half of the cases, annual growth is between 0 and 5%. In other places, growth exceeds 5%, or even 10%. These rates are significant. Indeed, an annual increase of 5% (and, respectively, 10%) means a doubling of traffic within 14 years (7 years respectively). These figures confirm the large increase in cycling traffic in major cities. The five count points that represent an exception can be explained by construction sites, route changes or even faults with the count points.

How do Swiss cities compare to other European cities? They are clearly below cities such as Houten, Amsterdam, Groningen in the Netherlands, Copenhagen in Denmark or Freiburg in Germany (Fig. 3.5). They are also outpaced by certain cities located in countries, which are typically less bicycle-friendly, such as Ferrara in Italy and Oxford in the UK. The modal share of Swiss cities, however, is higher than for many French, Mediterranean and British cities. Centres like Basel, Winterthur and Bern are even catching up to the Dutch cities with the lowest modal share in the country.

Cyclists' feelings of safety (see Sect. 10.3) and the assessment of how their needs are being taken into consideration by the public authorities (see Sect. 11.1) will clarify this point.

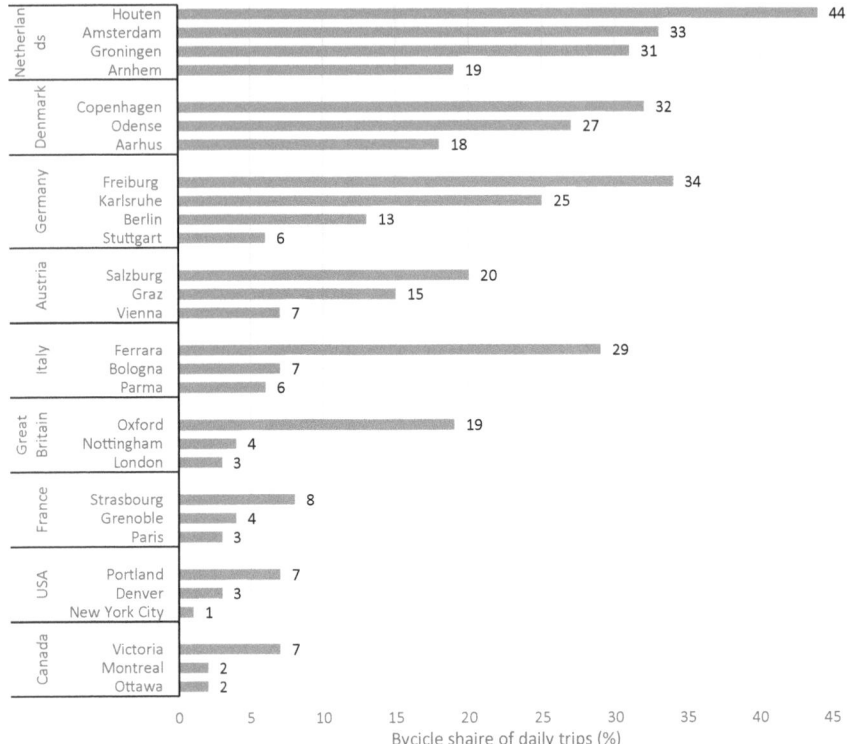

Fig. 3.5 Share of cycling in total journeys in various European cities (2010/2016). *Note* The data come from different sources (city mobility surveys, national statistical offices and European Commission). Methodological differences (data collection, date and schedule, spatial breakdown, etc.) limit the comparability of the data

3.3 Practices According to Profiles and Reasons for Travel

The Mobility and Transport Microcensus highlights differences between age groups and genders (Table 3.2). As these figures are based on samples, they should be

Table 3.2 Modal share of cycling in journeys according to gender and age, 2010–2015

	Gender		Age						Total (%)
	Men (%)	Women (%)	6–17 (%)	18–24 (%)	25–44 (%)	45–64 (%)	65–79 (%)	80 and above (%)	
2010	6.7	5.7	12.3	4.5	5.6	5.8	3.9	2.0	6.2
2015	7.3	6.4	12.1	6.0	6.5	6.6	4.6	2.5	6.8

Source Mobility and Transport Microcensus

Table 3.3 Modal share of cycling in journeys according to the reason for travel, 2010–2015

	Work (%)	Education (%)	Shopping (%)	Business travel (%)	Leisure (%)	Accompaniment (%)	Other motivations (%)	Total (%)
2010	7.2	12.5	5.5	4.6	5.7	2.1	3.9	6.2
2015	8.5	12.8	5.8	4.6	6.5	2.6	3.5	6.8

Source Mobility and Transport Microcensus

considered as orders of magnitude. Men use a bicycle in 7.3% of the journeys they make, which is a higher value than for women (6.4%).

The modal share of cycling is highest among minors aged 6–17 (12.1%), an age group that does not yet have access to a car (at least not independently) and which travels shorter distances (to get to a place of education, for example). It decreases by half for young adults aged 18–24 (6.0%) and then increases. After retirement, ageing decreases the practice of cycling. The values observed in 2015 are higher than those from 2010 for all age groups except the youngest, while a declining trend in cycling among children and adolescents has been observed for several years (see [5]).

In terms of reason, two are above the global modal share of cycling: education-related journeys (12.8%) and work-related journeys (8.5%), with the latter category experiencing the highest growth since 2010 (when its share was 7.2%). The mobility performed within the framework of leisure activities (as an activity in itself, or to reach a leisure destination) is slightly below the total. The lowest values relate to trips performed in the course of exercising a profession and for accompaniment journeys (e.g. taking a child to school) (Table 3.3).

The average journey length is 3.3 km for bikes without electric assistance and 4.4 km respectively for bikes with electric assistance [2]. The time taken to cover these distances is 16.5 and 15.5 min, respectively, at a speed of 13.3 and 17.0 km/h. The time taken on journeys made by bicycle varies significantly, however; this can largely be explained by the fact that a certain proportion of journeys are undertaken for leisure or sport, which are longer than average.

* * *

With almost 7% of journeys being made by bicycle, Switzerland is characterised by a higher modal share of cycling than that of the English-speaking and Latin countries, but lower than that observed in Northern Europe. This share, after having decreased, experienced a slight rebound between 2010 and 2015. Taking into consideration the type of space, it can be seen that the increased uptake of cycling primarily concerns large and medium-sized cities and certain municipalities on the outskirts of those cities, suggesting that cycling has become more urban in recent years. However, there are big differences: Basel and Winterthur are not far behind certain northern cities, with modal shares in excess of 15%; these are closely followed by Bern, Zurich and Lucerne, and far exceed other large population centres, such as Geneva, St. Gallen and Lausanne. The gap between Switzerland and certain European nations, as well as the recent growth observed in major Swiss cities, shows the potential of cycling in a country where 60% of journeys are shorter than 5 kms.

References

1. R. Buehler, Bicycling levels and trends in Western Europe and the USA. GeoAgenda **1**, 10–12 (2018)
2. OFS, & ARE, *Comportement de la population en matière de transports: Résultats de microre-censement mobilité et transports 2015* (Office fédéral de la statistique & Office fédéral du développement territorial, Neuchâtel & Berne, 2017)
3. P. Rérat, The new demographic growth of cities: the case of reurbanisation in Switzerland. Urban Stud. **49**(5), 1107–1125 (2012). https://doi.org/10.1177/0042098011408935
4. P. Rérat, The return of cities: the trajectory of Swiss cities from demographic loss to reurbanization. Euro. Plan. Stud. **27**(2), 355–376 (2019). https://doi.org/10.1080/09654313.2018.154 6832
5. D. Sauter, K. Wyss, *Etude pilote sur l'utilisation du vélo chez les jeunes dans le canton de Bâle-Ville* (Département des constructions et des transports & OFROU, Bâle & Berne, 2014)

Part II
The Participants in the Bike to Work Scheme

Chapter 4
Research Approach

What are the objectives of this research? How were bicycle commuters identified and approached? What are their characteristics?

This chapter describes the approach adopted in this research. First, we specify the objectives of the study. Next, we explain the *bike to work* scheme as well as the methods used to analyse the different dimensions of utility cycling. Finally, we discuss the profile of a sample of commuter cyclists, in terms of gender, age and socio-economic status.

4.1 Objectives of the Study

This research focuses on people who use bicycles as a means of transport to get to their place of work, whether on a regular basis or otherwise. This is based on the 2016 *bike to work* scheme. Participants in this national initiative are surveyed in order to gain in-depth insight into utility cycling, to analyse its main dimensions and to identify elements of tension arising as the cycling system is redefined.

This research operates on the principle that the results obtained from the *bike to work* participants provide a foundation for understanding utility cycling. Indeed, the experience of commuting is considered to be indicative of the use of bicycles as a means of transport. Furthermore, the mechanisms highlighted within this population may identify barriers to the wider adoption of the practice among populations, which are either less convinced or less competent.

Our examination addresses the components of the reading grid discussed in Chap. 2. The cycling system or velomobility refers to the three dimensions of uses, individuals' cycling potential, and a territory's hosting potential. The majority of the research focuses on the situation at the time of the survey. As an event deliberately created to promote cycling, *bike to work* may cause participants to make not only

P. Rérat, *Cycling to Work*,
SpringerBriefs in Applied Sciences and Technology,
https://doi.org/10.1007/978-3-030-62256-5_4

temporary but also long-term changes to their regular practices, and as a result, the analysis may at points adopt a temporal perspective by focusing on these impacts.

After characterising the profile of the *bike to work* participants at the end of this chapter, we focus on the journeys undertaken by bicycle (Chap. 5) according to frequency, distance and motivation. We also investigate the participants' use of other means of transport to identify the relative importance of the bicycle, and how it is combined with other travel options.

Regarding the participants' cycling potential, we focus on equipment (in terms of access to vehicles or membership of mobility schemes; Chap. 6), as well as their skills and level of ease with various types of coexistence with road traffic (Chap. 7). A substantial part of the analysis focuses on the factors underlying the appropriation of cycling as a means of transport. These factors include both motivations (Chap. 8) and barriers and disincentives (Chap. 9).

The territory's hosting potential is discussed in Chap. 10 from two perspectives: the spatial contexts in which utility cycling is most developed, and the participants' assessment of the bikeability of their commuting journeys. Finally, Chap. 11 focuses on the cyclists' perceptions of the extent to which cycling is taken into account by the public authorities in their region and of the measures put in place to promote bicycle commuting.

4.2   The Bike to Work Scheme

The *bike to work* scheme has been organised every year since 2005 under the aegis of PRO VELO Switzerland, the umbrella association representing cyclists' interests. It forms part of this association's global strategy to promote cycling. In 2016, the year in which the survey was conducted, the initiative brought together nearly 54,000 participants, divided into 14,000 teams from 1,800 companies. *Bike to work* is described on the dedicated website[1] as a 'Swiss-wide health promotion campaign' for companies. By participating, the latter 'strengthen their staff's team spirit and fitness, while supporting sustainable mobility practice'.

The initiative can be seen as a *nudge*, a measure intended to modify an individual's behaviour—in this case, to encourage them to try cycling—not through coercion but through incentivisation [5]. Mobility behaviour is often governed by habit, the most common, and most stable, form of mobility organisation [4], such that any potential modal shift may simply collide with and be prevented by the inertia of habitual mobility practices. Change is an onerous process, and information alone is not enough to cause people to break habits and adopt new behaviours. Causing them to *want* to change, however, and giving them the opportunity to change their pre-conceptions, is more likely to lead to a change in behaviour.

Campaigns to promote bicycle commuting exist in many countries, but in different ways [1, 2]. The Swiss *bike to work* campaign involves several stages. Any business

---

[1] www.biketowork.ch.

and institution—whether private or public—can register by making a modest financial contribution.[2] Teams—usually made up of four people—are then formed within these companies. During the months of May and/or June, each team undertakes to travel to work by bicycle as often as possible. Participants record the kilometres they have travelled in an activity calendar, and those who have completed at least half of their commutes by bicycle are entered into a competition to win a variety of prizes.

This scheme offers numerous areas of interest given the objectives of this research. It provides us with a large sample, covering a range of profiles. The initiative brings together experienced cyclists (who already complete all of their home–work journeys by bicycle), occasional cyclists (who take advantage of the *bike to work* event to do more cycling), and less experienced cyclists (who, motivated by their colleagues, embrace the opportunity to give bicycle commuting a try).

4.3   The Field Survey

A questionnaire drafted in both German and French was developed to address all the dimensions discussed above. In September 2016, the organisers of the *bike to work* scheme sent an email inviting participants to complete the online questionnaire. Out of a total of nearly 54,000 participants, 44,726 emails were sent. The reason for this discrepancy is that not all team leaders provided their teammates' contact details; this is predominantly the case for people who do not use a personal computer for work (for example, sales or craft workers), or who do not have a professional email address. It was not possible to send a reminder message. A total of 14,620 questionnaires were collected. After excluding any forms that were not sufficiently filled out, 13,744 were retained, which represents a response rate of 30.7%.

The data were then tested and cleaned before being used as a basis for statistical analysis. The analysis did not include those participants (3.1%) who used another active mode (walking, scooter, etc.) as part of the scheme, while still wishing to promote cycling, such as this person:

> It is precisely because I don't like to travel by bicycle in urban traffic that I participated in the bike to work *scheme as a pedestrian. I feel safe on the pavement but less safe on the road.*

Throughout this book, the totals in the charts and tables may vary. Indeed, the questions in the questionnaire were not made mandatory so as to leave the participants free to answer or not as they so wished. However, the differences observed are minimal compared to the large volume of the sample.

The data from the questionnaire were used to quantify the various dimensions of cycling. Respondents were also given the opportunity to leave comments in multiple places. Many of them did so, such that around 11,000 comments were collected. Interviews were also conducted with 30 participants who study or work at the University

---

[2]The amount is 100 Euros for companies with fewer than ten people. It then increases non-proportionally according to the number of employees.

of Lausanne. The comments and interview transcripts were analysed in order to supplement the quantitative results and clarify interpretation. The quotes, using the respondents' own words, illustrate the different situations, choice processes, feelings and experiences of the participants in the *bike to work* scheme.

4.4   Profile of the Sample

Using the *bike to work* scheme enabled us to identify a very large sample, bringing together a variety of practices; however, the studied population has certain specificities.

First, it excludes—by its very nature—people who cycle exclusively for reasons other than commuting, including shopping, leisure, sport, etc. Second, it does not include any retired people, or those who are not in paid employment. Most of the participants are economically active, although there are a small number of university students (1.7% of the sample). Finally, it is more difficult for self-employed workers and employees of small companies to form teams. Likewise, part-time workers may be less likely to take part in the scheme.

Comparisons can be made to identify certain characteristics of the sample in terms of age, gender and level of education (Table 4.1). Two sources were used—the Swiss Active Population Survey (2016) and the Mobility and Transport Microcensus (2015)—with the aim of comparing the sample against, firstly, the wider economically active population and, secondly, cyclists, defined here as people who used a bicycle on the reference day of the Mobility and Transport Microcensus.

Table 4.1 Comparison of the *bike to work* participants sample against the economically active population in Switzerland and the Mobility and Transport Microcensus

Variables	Modalities	*Bike to work* survey (2016) (%)	Active population (2016) (%)	Bicycle users (2015) (%)
Age	15–24 years	3.9	12.5	15.4
	25–39 years	35.6	32.2	26.2
	40–54 years	46.0	35.3	31.6
	Over 55 years	14.5	20.0	26.8
Gender	Women	42.0	46.6	45.3
	Men	58.0	53.4	54.7
Education	Compulsory education	1.5	12.6	10.5
	Initial and higher vocational training, high school diploma	44.5	46.2	51.5
	Tertiary colleges	54.0	41.2	38

Source Questionnaire, Swiss Active Population Survey and Mobility and Transport Microcensus

Two age groups are over-represented among *bike to work* participants—25–39 year olds and 40–54 year olds—while the youngest and oldest age groups are less present. In terms of type of household, almost half live in a household with one or more children, and more than one in four lives with a partner; the other respondents live in non-family households (people living alone, in a shared residence or with their parents).

In terms of gender, women are somewhat under-represented. An initial explanation for this difference lies in their lower participation in the labour market (only 46.6% of the working population are women) and this is particularly true in the age groups which are over-represented in the sample (due to withdrawal from the labour market or reduction in working hours in order to raise a family).[3] Another explanatory factor is the slightly less frequent use of bicycles among women in general; women represent 45.3% of people who used a bicycle the day before the Mobility and Transport Microcensus. However, in our survey, there are marked differences between cantons. There is evidence of a phenomenon observed in the literature (see 2.4.2), namely that the higher the modal share of cycling, the more women are represented among bicycle users. Women are thus in the majority (56.9%) in the cyclophiliac canton of Basel-Stadt, but constitute only a third of participants in cantons where cycling is practised less frequently (Vaud, Tessin, Neuchâtel, Valais, etc.).[4]

In terms of level of education, people who completed the compulsory level of education and then left school are under-represented in the sample compared with the wider active population, unlike those with a higher education. This trend does not seem to be due to the increased popularity of cycling among highly qualified people, as they are less numerous among bicycle users in general than in the active population. Instead, the over-representation of highly skilled people in the sample may be explained by a higher awareness of the arguments relating to health, a higher need to integrate physical activity into the working day as they are engaged in non-manual activities, an over-representation of large companies and institutions and certain fields of activity, etc. With regards to this last point, the four most represented sectors are public administration (19.3%), industry (chemical and pharmaceutical, food; 18.8%), health and the social sector (18.2%), and teaching and research (14.6%).

The particularities of this sample—people of working age, over-representation of 25- to 54-year olds, holders of a university degree, those working in companies within the upper tertiary sector—must be taken into account in the analysis. We will point out any results, which may differ to data on other utility cyclists. However, it is important to emphasise that the sample obtained is heterogeneous in many dimensions, and covers a multitude of profiles. Moreover, for the majority of the points addressed in the research—skills with regard to different traffic situations, barriers, experiences, etc.—the composition of the sample has little impact. The

---

[3]According to the Structural Survey, the shares of men and women commuting by bike in Switzerland are actually similar and both account for 7% [3].

[4]In general, we identify an over-representation of participants in urban and German-speaking cantons. The spatial distribution of participants in the *bike to work* scheme is broken down in Sect. 10.1.

results provide relevant and generalisable insights into the practice of utility cycling in Switzerland.

* * *

Each year, the *bike to work* scheme brings together thousands of commuters in Switzerland who commit to using bicycles for all or part of their commute, which has made it possible to identify people who engage in utility cycling across the country. During the 2016 event, nearly 14,000 participants responded to a questionnaire, thus creating an extensive database of information.

In comparison with the wider Swiss economically active population, the profile of *bike to work* participants is characterised, in particular, by an over-representation of people from the middle age categories and of those who hold a university or equivalent level of education and are working in the upper tertiary sector. This difference with the wider population can be explained by differing propensities to engage in utility cycling and, particularly, to take part in the *bike to work* scheme (depending on the number of employees, the size and type of business, and interest in the initiative and the arguments used). The following chapters will show a diverse sample in terms of mobility practices and the appropriation of cycling.

References

1. D.J. Lee, Embodied bicycle commuters in a car world. Soc. Cult. Geogr. **17**(3), 402–420 (2015). https://doi.org/10.1080/14649365.2015.1077265
2. D. Piatkowski, R. Bronson, W. Marshall, K.J. Krizek, Measuring the impacts of bike-to-work day events and identifying barriers to increased commuter cycling. J. Urban Plan. Develop. **141**(4), 04014034 (2015). https://doi.org/10.1061/(ASCE)UP.1943-5444.0000239
3. OFS, *La pendularité en Suisse 2016* (Office fédéral de la statistique, Neuchâtel, 2018)
4. J. Scheiner, C. Holz-Rau, Changes in travel mode use after residential relocation: a contribution to mobility biographies. Transportation **40**(2), 431–458 (2013). https://doi.org/10.1007/s11116-012-9417-6
5. R.H. Thaler, C.R. Sunstein, *Nudge: Improving Decisions About Health, Wealth, and Happiness* (Penguin Books, New York, 2009)

Chapter 5
Journeys

For what types of commuting journeys are bicycles used? How does their use vary within the studied population? What is their place within mobility practices?

This chapter focuses on the uses of bicycles. It identifies the main characteristics—frequencies, distances, patterns—of the journeys undertaken by bicycle. A typology of cyclists is proposed based on these characteristics, and the use of other means of transport, and the way in which they are combined with cycling, is also documented.

5.1 Participants Are Mainly Cycling Enthusiasts

In 2016, a third of respondents took part in the *bike to work* scheme for the first time. For half of the respondents, it was at least their third time taking part. The majority of participants are regular bicycle commuters (Fig. 5.1): three quarters of them already used bicycles regularly (17.2%) or most of the time (56.2%) in their commute before *bike to work*. Fewer than 1 in 10 started to ride their bicycles to work as part of this event. Despite representing a minority of participants, the number of new cyclists amounts to several thousand individuals each year. This shows the significant potential of the event in terms of promoting cycling.

Taking part in *bike to work* has a learning and training effect that can encourage changes in habits. Three months after the event, 65% of participants who had not previously been in the habit of using a bicycle to travel to work stated that they were now cycling more, while 50% of people who previously used a bicycle on an irregular basis said the same. The effect is not always immediate, as several participants point out:

> It's mid-September and I'm still cycling. After my first time taking part in bike to work *in 2015, there was no lasting effect and I went straight back to using the car.*

> My 1st *time taking part, I had a standard bicycle. I told myself that I would never take part again, as my journey was much too long (35-40 minutes). In the 2nd and 3rd years, I used*

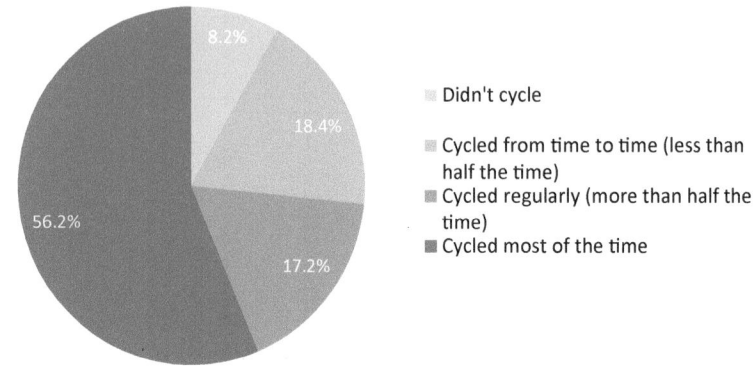

Fig. 5.1 Use of cycling to get to work before participating in *bike to work*

a 25km/h electric bike. It was OK, but there was no particular enjoyment. So I only did June, then nothing more for the rest of the year. In the 4th and 5th years, I used a [45 km/h electrically assisted bicycle]. Travelling has really become a pleasure. Without bike to work, I would never have taken the step.

5.2   Characteristics of the Commuting Journeys

While a large majority of respondents (85%) use bicycles for their entire commute (Table 5.1), a significant minority combines cycling with public transport. In this case, cycling plays a more significant role (11.2%) in what is known as the 'first kilometre', i.e. from home to the station, than in the 'last kilometre', i.e. from a public transport stop to the final destination (4.4%). 2.6% of participants take their bicycles on public transport. The total of the responses exceeds 100%. Indeed, some commuters use two bicycles, i.e. one before and a second after using public transport. Others modify their behaviour depending on the day or the season (for example, they may cycle for the whole journeys during nice weather and only for part of the journey in winter).

Intermodality—i.e. combining cycling with another mode of transport within the same journey—therefore represents a feature of a significant proportion of journeys,

Table 5.1 Combination of cycling and public transport in commuting journeys

When travelling to work, you cycle…	Percentage (%)
From home to the workplace	85.0
Before taking public transport	11.2
After taking public transport	4.4
By taking it on public transport	2.6

Note The total is more than 100%, as multiple answers were possible

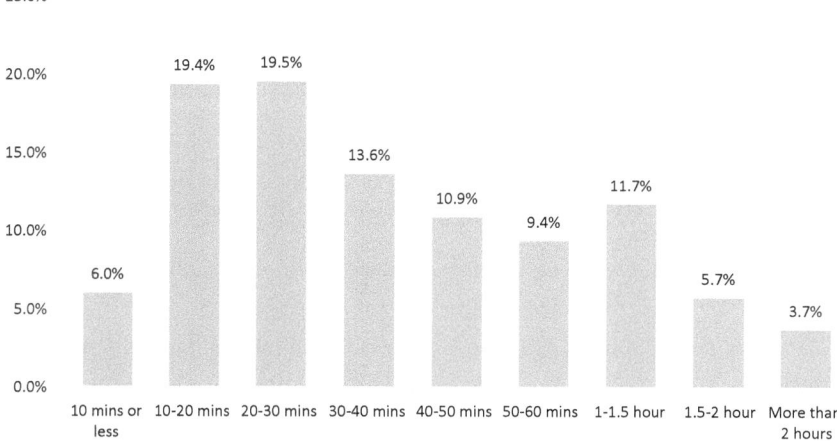

Fig. 5.2 Round trip travel time from home to work on a conventional bicycle

and is likely to increase. In the context of longer commuting journeys, bicycles can be attractive due to their complementarity with public transport. First, they allow for integration with railways, as shown by the development of different forms of parking (sheltered or in the open, secured or not, controlled or not, etc.).

The length of the commuting journey is expressed in number of minutes taken to complete a round trip (Fig. 5.2). These journey times only relate to cycling and exclude any other modes used during the journey. On average, a participant in the *bike to work* scheme spends 47.8 min per day on their bicycle if it is mechanical, and 53.2 min if it is electrically assisted.

For two-fifths of participants, the cycled journey lasts between 10 and 20 min, and it lasts between 20 and 30 min for a similar proportion. Round trips lasting less than 10 min are rare, as cycling is in competition with walking at this distance. The duration of the journeys is long or very long for the rest of the sample. Almost one in five cyclists spends more than an hour on their bicycle per day.

The home–work distances were estimated based on origins and destinations (as-the-crow-flies distances were calculated using postcodes) and stated travel times. In our sample, the average distance is estimated at 5.9 kms for users of a conventional bicycle. This value represents almost double that measured by the Mobility and Transport Microcensus for commuting journeys completed using a mechanical bicycle (3.3 kms in 17 min). It is also greater than the journeys undertaken using an electrically assisted bicycle (4.1 kms in 16 min for electrically assisted bicycles capped at 25 km/h and 5.4 kms in 15 min for electrically assisted bicycles capped at 45 km/h).

These differences are explained by several characteristics of the *bike to work* scheme. While it attracts some who are new to cycling, it also motivates sports cyclists to use their bicycles for utility purposes, provides a challenge that is limited to the months of May and/or June and encourages competition between teams and teammates. This is illustrated by the following testimonies:

> *For sporting reasons, for* bike to work, *I increased the distance and the vertical metres on my return journey.*
>
> *I even often do an extra circuit when I get home.*
>
> *During* bike to work, *I do a much longer bike ride than during the rest of the year. The effort is greater than usual. I leave more time and I bring different types of additional clothes (rain protection, spare clothes).*
>
> *During* bike to work, *bicycles are the 'normal' means of transport; at all other times, the opposite is true.*

These last two comments raise the issue of seasonality. Unlike users of cars or public transport, cyclists are in direct contact with the weather conditions. Cycling can therefore be seasonal in nature due to temperature, precipitation, light levels, road conditions (lack of snow clearance or streetlights, especially on country roads), etc. However, the latter varies greatly depending on the country; the more established the practice of cycling, the less seasonal it is (see Sect. 2.4).

The impact of weather conditions also differs between individuals (as we will also see when we discuss barriers in Sect. 9). In our sample, cycling is constant over the year for 4 out of 10 cyclists, while it is more frequent in the summer for a third of participants. For less than a quarter, the practice is markedly seasonal in nature; these people state that they only cycle in summer.

The seasonal nature of cycling can be explained, in particular, by the length of journey. Those who adopt stable cycling behaviour throughout the year require an average of 36 min per round trip for their commuting journeys; this value increases to 49 min for those who state that they cycle less in winter and 63 min for those who limit their practice of cycling to the summer months.[1]

Another explanatory factor is the relatively high level of experience of many participants, which allows them to adopt non-seasonal behaviour. For others, competition between team members during the *bike to work* scheme encouraged them to cycle even in rainy weather and to acquire new habits. They either realised that it is not prohibitive or acquired appropriate equipment (Fig. 5.3):

> Bike to work *motivated me to cycle even in the rain. I got myself some good clothes and I noticed that it wasn't a problem at all.*
>
> *I am more likely to choose to cycle even when the forecast is for rain. experience has shown me that, even when it does rain, it doesn't rain much or for long!*
>
> *I've started using the bike in rainy weather. I didn't do this at all before.*
>
> *I use weather less as an excuse than I did previously...*

---

[1]The estimated distances are respectively 9.1, 12.3 and 16.0 kms.

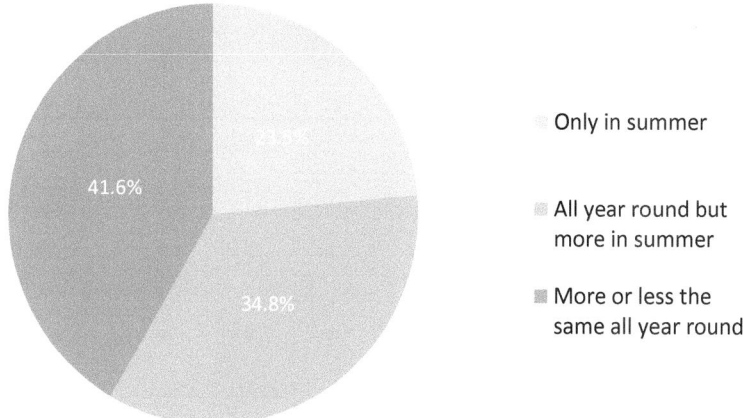

Fig. 5.3   Seasonal use of the bicycle to travel to work

5.3   The Place of Cycling in Mobility Practices

The results relating to the length of bicycle journeys and their sometimes seasonal nature lead us to consider the reasons for travelling by bike (Fig. 5.5) and the way in which it connects with other modes of travel as part of the home–work commute (Fig. 5.4).

In the sample, the bicycle is the most common means of transport for getting to work: almost 60% of participants say that they cycle for this purpose the majority of the time (Fig. 5.5). The others exhibit multimodal behaviour by using other modes with varying degrees of regularity (Fig. 5.4). The respondents prefer public transport

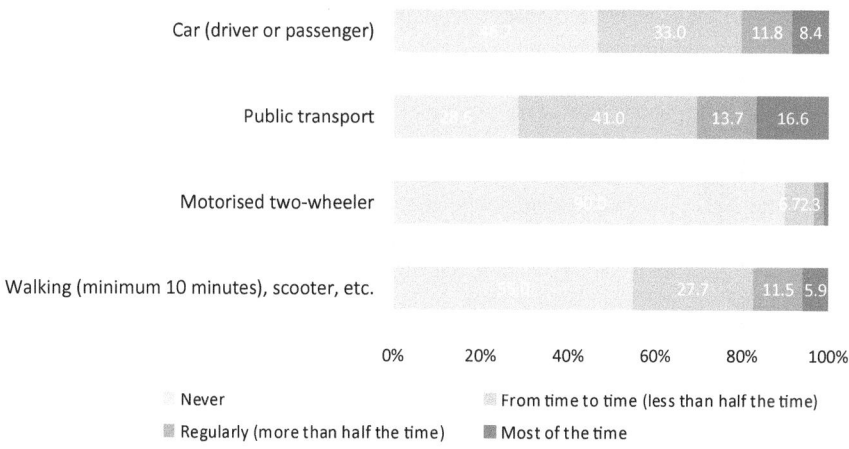

Fig. 5.4   Frequency of use of modes of transport other than cycling to travel to work

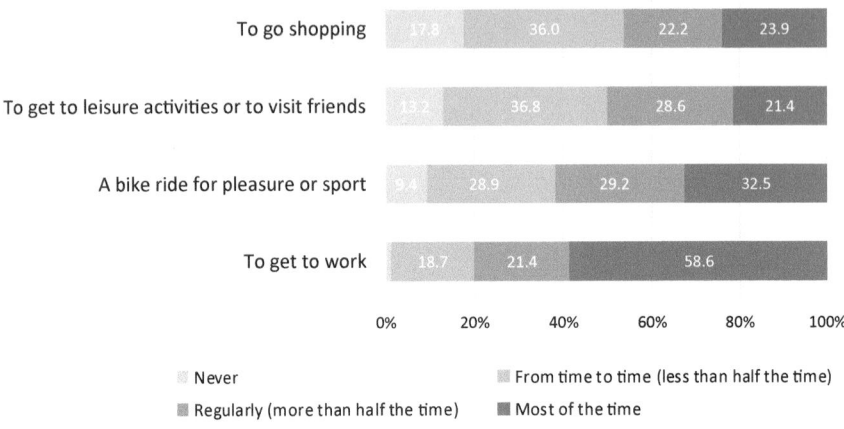

Fig. 5.5 Frequency of bicycle use according to reason for travel

when they do not cycle to work, and more than 30% use public transport regularly or the majority of the time. The proportion of participants who use a car is lower: almost one in two never uses one, either as a driver or a passenger. Finally, 17% walk regularly or frequently and 3% use motorised two-wheelers.

Among the participants, bicycles are therefore used more frequently than other means of transport for commuting. Similarly, commuting is the most common reason for cycling (Fig. 5.5): 80% of the respondents regularly or always commute by bicycle. A few percentage points separate the frequencies observed here from those prior to *bike to work* (Fig. 5.1), which confirms the change in behaviour and adoption of cycling by some novice cyclists.

The practice is not limited to commuting. Around half of the participants state that they use a bicycle for other purposes, such as going out, seeing friends, shopping, etc. 60% also use bicycles for pleasure or as a sport. In this case, the boundary between utility cycling, leisure cycling and sports cycling is permeable. For the other 40%, cycling is above all—if not exclusively—a mode of travel.

By considering both how often people travel by bike, and why, we can develop a typology made up of six types of cyclist: systematic, leisure, individual, utility, commuter and occasional cyclists (Table 5.2).[2] These categories are cross-referenced against the cyclists' profiles (gender, type of household)[3] and the commuting journeys they undertake, in order to interpret the specificities (Table 5.3).

The first category is that of systematic cyclists (15% of the sample). These cyclists use bicycles frequently or even exclusively, both for utility purposes (shopping, going

---

[2]This typology is based on four frequency levels—never, from time to time (less than one in two journeys), regularly (more than one in two journeys), most of the time—and four travel motivations—going shopping, travelling to leisure activities, pleasure/sport, and going to work. The method used is the *two steps* method, which results in a division into six groups.

[3]There are no marked socio-economic differences.

Table 5.2 Typology of *bike to work* participants according to their bicycle usage

		Systematic cyclists (%)	Leisure cyclists (%)	Individual cyclists (%)	Utility cyclists (%)	Commuter cyclists (%)	Occasional cyclists (%)	Sample total (%)
Share (%)		**15**	**11.7**	**17.3**	**21.8**	**13.9**	**20.3**	**100**
Going shopping	Never or occasionally	0	36	**74**	17	**98**	**96**	54
	Regularly or most of the time	**100**	**64**	26	**83**	2	4	46
Travelling to leisure activities	Never or occasionally	4	28	**56**	18	**94**	**95**	50
	Regularly or most of the time	**96**	**72**	44	**82**	6	5	50
Pleasure or sport	Never or occasionally	0	5	0	76	**98**	38	38
	Regularly or most of the time	**100**	**95**	**100**	24	2	**62**	62
Going to work	Never or occasionally	0	**40**	0	1	0	**74**	30
	Regularly or most of the time	**100**	60	**100**	**99**	**100**	26	70

Table 5.3 Characteristics of types of cyclists according to their bicycle usage

		Systematic cyclists (%)	Leisure cyclists (%)	Individual cyclists (%)	Utility cyclists (%)	Commuter cyclists (%)	Occasional cyclists (%)	Total (%)
Profile								
Gender	Women	50.1	38.8	29.4	52.0	40.1	38.9	**41.9**
	Men	49.9	61.2	70.6	48.0	59.9	61.1	**58.1**
Residential context	Urban	57.5	36.2	32.1	53.6	33.5	22.7	**39.5**
	Suburban	35.4	44.2	50.8	38.6	51.2	51.1	**45.2**
	Rural	7.1	19.6	17.2	7.8	15.3	26.2	**15.3**
Household	Without children	55.7	53.1	42.7	54.0	44.5	51.3	**50.3**
	With child (ren)	44.3	46.9	57.3	46.0	55.5	48.7	**49.7**
Journeys by bicycle								
Seasonality of bicycle use	In summer	1.8	37.7	5.8	5.5	20.7	67.6	**23.4**
	All year	98.2	62.3	94.2	94.5	79.3	32.4	**76.6**
Commuting journeys by car	Occasionally	98.1	69.7	91.9	94.7	81.2	45.4	**79.9**
	Regularly or most of the time	1.9	30.3	8.1	5.3	18.8	54.6	**20.1**
Commuting journeys by public transport	Occasionally	79.8	49.9	79.4	75.0	69.0	59.9	**69.7**
	Regularly or most of the time	20.2	50.1	20.6	25.0	31.0	40.1	**30.3**
Commuting journeys	Duration of return journey	37'	62'	43'	35'	42'	65'	**47'**

to work, travelling to leisure activities) and when the journey is an objective in itself (for pleasure or sport).

This class is characterised by a higher proportion of women, people without children and city dwellers. The commuting journeys are on average shorter than for the sample as a whole (37 vs. 47 min) and are not seasonal in nature. Only 1.8% state that they only use bicycles in summer (compared to 23.4% in the wider sample).

The second category represents just over 1 in 10 participants: leisure cyclists. They are named thus due to the fact that they use the bicycle more regularly for free time activities (leisure, sport, travelling to other leisure activities) than for more compulsory activities (shopping, work).

The profile of this category is broadly similar to the sample as a whole (although with slightly fewer couples with children, and women). The journey time to reach the workplace is significantly higher than average (62 min vs. 47), and 37.7% of people in this category only cycle in summer. At other times of the year, they mainly use public transport, as well as the car, for commuting.

The next type is individual cyclists (17.3%). This group's use of bicycles is only partial and relates to journeys that are more easily completed alone, such as going to work or going for a bike ride. Conversely, their use of bicycles to go shopping or to travel to leisure activities is more occasional. These are activities that are most likely to be carried out by or with other members of the household.

The individual cyclist category is characterised by a marked under-representation of women (29.4%) and a larger proportion of people living in couples with children (57.3 vs. 49.7%). However, while their use of the bicycle is only partial, it takes place throughout the year.

Utility cyclists represent the largest category (21.8%). They use bicycles regularly, sometimes systematically, for all utility purposes. In this category, bicycles are perceived above all else as a vehicle, and cycling is less likely to be seen as a leisure activity.

This group has a predominance of women (52%, the maximum proportion) and a significantly larger proportion of city dwellers (53.6 vs. 39.5%). Like regular cyclists, their commuting journeys are shorter than the average and this category rarely uses cars.

Commuter cyclists (13.9%) mainly confine their use of the bicycle to journeys between home and the workplace, using it for few other aspects of daily life. The frequency of other reasons to use the bike is very low.

Among commuter cyclists, we can observe an over-representation of people without children (55.5 vs 49.7%). These characteristics are shared by leisure cyclists, which seems to suggest that the modal choice for certain activities results from needs and trade-offs not at the individual level but at the household level (commuting is a more individual choice, while other activities are more likely to involve the other members of the household).

Finally, we can identify the occasional cyclists (20.3%). This category is distinguished by a lower frequency of bicycle use for commuting. However, it is similar to commuter cyclists in terms of the low values associated with utility motivations or daily activities (travel for shopping and leisure purposes).

This group differs from the previous group by engaging more frequently in cycling as a leisure or sporting activity. Occasional cyclists are less likely than average to live in cities (22.7%) but are more numerous in suburban (51.1%) and rural areas (26.2%). They are therefore one of the only two categories—along with leisure cyclists—to take longer than average to get to work (65 vs. 47 min for a round trip) and to use bicycles only in summer (67.7 vs. 23.5%). Outside this period, they predominantly use cars.

* * *

Cycle trips are sometimes considered to be journeys over a limited distance, undertaken in good weather. In reality, they are actually much more diverse within the studied population. Less than a quarter of the participants state that they only cycle in summer, and the average distance travelled is close to 6 km (without electric assistance). While the specific nature of the *bike to work* scheme must be noted, this distance is higher than the threshold of 3–5 kms often used in planning matters in Switzerland to estimate the range of the conventional bicycle.

A typology of six categories of cyclist—systematic, leisure, individual, utility, commuter and occasional cyclists—shows variations in cycling practices in terms of frequencies and patterns. The extent to which cycling is used for utilitarian purposes varies for each type, and this utilitarian use of the bicycle interacts differently in each case with its use for leisure or sport. Each category of cyclists varies according to gender, life course position and residential location. Cycling occupies a significant place in the mobility habits of the majority of respondents.

For the majority of the participants, the bike has an important place in their mobility practices. Six in ten cyclists report that they cycle most of the time to get to work while for others, the most common alternative is public transport, followed by the car. It thus appears that the majority of cyclists exhibit multimodal behaviours, in the sense that they use various means of transport. The next chapter, which addresses the access to mobility equipment, will clarify this point.

Part III
Individuals' Cycling Potential

Chapter 6
Access

What types of bicycle do the *bike to work* participants have? To what other means of transportation do they have access?

The mobility portfolio or equipment of the *bike to work* participants is discussed in two stages, looking first at which type of bicycle they have and then at their access to or ownership of motorised vehicles and public transport passes. This section enables us to clarify the results of the previous chapter, in particular with regards to combining cycling and other modes of transport.

6.1 The Emergence of the Electrically Assisted Bicycle

Mechanical bicycles are, unsurprisingly, the best known and most widespread type of bicycle: almost 9 out of 10 participants have one and more than 80% used it for *bike to work* (Table 6.1). However, the electrically assisted bicycle is also on the rise, as indicated by the fact that 18% of participants own one, even if not all of them used it during the scheme.[1]

Electrically assisted bicycles play a role in extending the practice of cycling across profiles and territorial contexts. Women (49.6%) are almost as likely as men (50.4%) to use electric bikes, while the gap is larger for conventional bicycles (41% versus 59%). E-bike uses are also older: 76% are aged over 40, a much higher percentage than for conventional bicycles (57.4%). The electrical assistance allows more women and people in the second half of their professional career to continue to cycle while exerting less physical effort. The lowest incomes (less than 3,000 francs

---

[1] A clear majority (67.2%) of the e-bikers also own a mechanical bike. Their practice of regular cycling may decline and even be taken over by the e-bike. Finally, 32.8% of e-bike users do not have a mechanical bike. The survey does not provide information about how many of them gave up their mechanical bike or started cycling again after purchasing an e-bike.

© The Author(s), under exclusive license to Springer Nature Switzerland AG 2021
P. Rérat, *Cycling to Work*,
SpringerBriefs in Applied Sciences and Technology,
https://doi.org/10.1007/978-3-030-62256-5_6

Table 6.1 Types of bicycle owned and used to participate in *bike to work*

	Type of bicycle owned by participants (in %)	Main type of bicycle used for *bike to work* (in %)
Conventional bicycle (road, city, mountain bike)	89.8	82.6
Electrically assisted bicycle	18.4	16.3
Folding bike	1.6	0.4
Other bicycle (cargo bike, recumbent bike, etc.)	2.6	0.5
Bike share scheme (membership)	0.6	0.1
None	0.2	–

Note Since respondents may own more than one bike, the total is greater than 100%

per month) are also slightly over-represented (4.9% versus 3.6%). In this case, electrically assisted bicycles may replace either a motorised vehicle or a public transport pass.

There are also noticeably fewer residents of urban centres among e-bike users then among those who own a conventional bicycle (23.1% versus 42.9%) Instead, e-bike owners include more inhabitants of the suburbs and rural areas. These vehicles make it possible to travel longer distances. As we saw in the previous chapter, a round trip takes 53 min on average for electrically assisted bicycles, versus 48 min for conventional bicycles, and is completed at a higher speed. In a territorial context marked by urban sprawl, electrically assisted bicycles thus represent an interesting way in to promoting cycling and, in particular, encouraging this practice among populations that leave the city to settle in the suburbs.

Other types of bike—foldable, cargo, recumbent, etc.—are uncommon and their usage rate is lower than the rate of ownership. Membership of bike share schemes is lower still (0.6%), and bicycle sharing networks in existence at the time of the survey remain small and sparse. While they provide a complementary offer, they are unable to compete with or replace ownership of a bicycle. But with the development of large-scale systems of more than 2,000 units in Bern and Zurich, bike sharing is set to increase in the future. Finally, a very small proportion of participants did not own a bicycle (0.2%). In this case, they benefited from a bicycle that had been loaned to them by colleagues or their company. The experience, if it proved positive, encouraged them to purchase their own equipment:

> I bought an electrically assisted bicycle after [having participated in *bike to work*] and I use it every day now, because I can do the big climb on the way back and I no longer have to fall back on public transport.

According to the comments left in the questionnaire, *bike to work* more commonly has an impact on equipment before the scheme actually begins. Some participants clarified that the initiative had prompted them to repair their existing bicycle (*'Dust off the bike!'*), service it or acquire appropriate equipment for commuting:

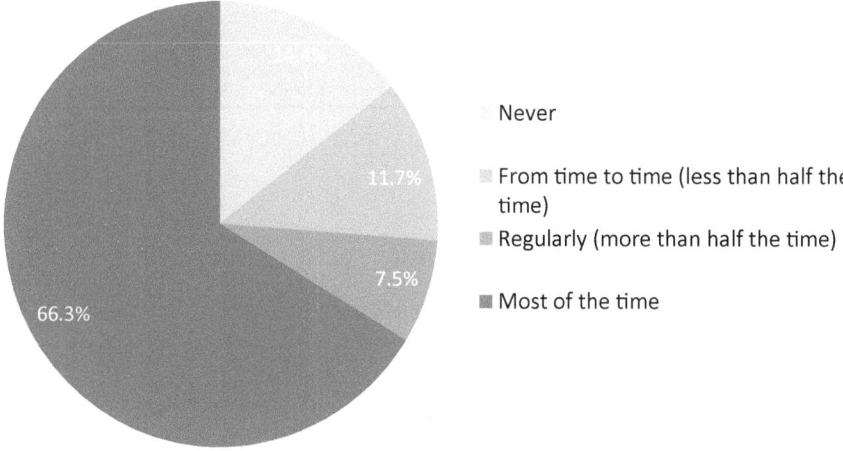

Fig. 6.1 Frequency of using a helmet

It gave me the impetus to fix my bike. that's why I used it more often afterwards.

I bought an electric bike to be able to keep my commitment, because I no longer have the physical fitness for a normal bike.

Because of bike to work, I bought a cycling helmet and a padlock, and I got my old bike back in working order by servicing it. I have discovered that it is quite possible to travel to work by bicycle.

Another factor relating to equipment is the wearing of a helmet, which is commonplace among the surveyed population. Two-thirds say they wear one most of the time; this value is even higher among users of electrically assisted bicycles (86.5% versus 62.4% for users of conventional bicycles) due to their higher speed and the legal obligations in respect of bicycles that provide electric assistance up to 45 km/h (Fig. 6.1).

6.2 Access to Other Means of Transport

The level of ownership of motor vehicles and public transport passes was compared against the results of the 2015 Mobility and Transport Microcensus to highlight the specificities of the sample of cyclists interviewed.

Almost all of the *bike to work* participants have a driving licence (95.3%), which shows that they have been socialised to motorised mobility. The proportion of participants who always have a private car available to them is just over 50%, this value is much lower than that observed at the national level (74% of economically active people aged between 18 and 65). The sample nevertheless shows an over-representation of people who have access to a car on demand (27.9% versus 20%), but also of those who do not have access to a private car (20.1% versus 6%).

Table 6.2 Access to a private car and full public transport pass

		Access to private car			Total (%)
		Yes, always (%)	Yes, on demand (%)	No (%)	
Public transport passes (excluding half-fare card)	Yes	11.9	10.1	10.5	32.6
	No	40.0	17.8	9.6	67.4
Total		51.9	27.9	20.1	100

This is offset by a higher proportion of members of car-sharing organisations (17% versus 4%). 13.7% also own a motorised two-wheeler (11.7% in Swiss households in general).

The lower level of car ownership is explained by the fact that the participants in the *bike to work* scheme belong to a more urban population, and that many of them are able to do without a car thanks to their frequent use of cycling. However, the majority still use cars on an ad hoc or even regular basis.

This population is well equipped in terms of public transport passes. The sample is more likely to hold a general travel pass or a half-price fare card (16.7% and 61.9%) than the wider Swiss active population (10% and 35%).[2] These figures are in addition to regional (12.2%) or route-specific (4.9%) travel passes, such that only 16.9% have no travel pass at all.

Cross-referencing the availability of a private car and that of a full public transport pass (without additional user expenditure per journey, thus excluding half-fare cards) enables us to identify nearly 10% of people who exclusively cycle in the pursuit of their regular activities (no car, no travel pass). 18% mainly use a bicycle for commuting, as they have no travel pass or permanent/on-demand access to a private car (Table 6.2).

Conventional cyclists rely more on public transport and are more likely to have a national or a regional public transport pass than e-bikers (17.9% and 12.8% respectively versus 10% and 9.1%). Although the great majority has a driving licence in both cases, e-bikers are much more motorised: 63.6% always have a car at their disposal and only 8.8% do not own a car, while less than half the conventional cyclists always have access to a car and a quarter of them never do. The difference here is not due to the kind of bike per se but rather to the fact that the e-bike appeals more to people who are more motorised than average: those older than 40, with children, in suburban and rural areas (for the place of residence, see Sect. 10.1).

* * *

Almost 9 in 10 cyclists have a conventional bicycle. The number with access to electrically assisted bicycles—almost one in five participants—reflects the steady

---

[2]The general travel pass provides access, in particular, to the railways and Post Bus networks, as well as to urban public transport throughout Switzerland. The half-fare card gives a 50% discount on train tickets.

growth in popularity of this type of vehicle in Switzerland. The audience for the e-bike is more female, older and less urban than for the conventional bike. It is more prevalent in the suburbs and rural municipalities and is used over longer distances. Electric assistance thus makes it possible to broaden the practice of cycling both in terms of population and type of space.

Almost all the cyclists are able to drive (95% of them have a driving licence). Although half of them still have access to a private car, their motorisation rate is significantly lower than that of the Swiss working population as a whole. The proportion with public transport passes is much higher, and the majority of participants appear to be multimodal (with various combinations of modes of transport), while 10% can be classed as exclusive cyclists (without access to a car and without a general or regional travel pass).

Chapter 7
Skills

We are all familiar with the saying, 'You never forget how to ride a bike!'—but what are the specific skills that utility cyclists need? How does their level of ease vary according to the infrastructure and to the degree of cohabitation with car traffic?

Getting around requires skills that are often overlooked in the study of mobility and transportation. These skills must allow a person to get around within different territorial contexts and for the purposes of carrying out their activities. We explore how the level of ease with which cyclists can achieve this varies according to the type of route, the degree of coexistence with motor vehicles, and a number of physical, technical and organisational skills. The way the cyclists behave in traffic and the impact of the *bike to work* experience on their levels of ease are also analysed.

7.1 Levels of Ease Depending on Degree of Cohabitation with Road Traffic

Riding a bike involves a certain mastery of balance and pedalling, and this skill has obviously been acquired by the population under study. Utility cycling, however, requires additional skills: it necessitates the ability to get around in different situations, which we approach from the perspective of levels of ease. A low level of ease will result in a cyclist making detours to choose more suitable routes, feelings of unease, or even the rejection of cycling altogether, depending on the context.

In general, the cyclists' levels of ease vary greatly depending on the route and the degree of coexistence with road traffic (Fig. 7.1). The maximum score is achieved for cycle tracks, which are, by definition, physically separated from road traffic: 95% of respondents say that they are (very) at ease on these. The remaining 5% have had experiences where cycle routes have turned out to be spaces which are shared with pedestrians and other users:

© The Author(s), under exclusive license to Springer Nature Switzerland AG 2021
P. Rérat, *Cycling to Work*,
SpringerBriefs in Applied Sciences and Technology,
https://doi.org/10.1007/978-3-030-62256-5_7

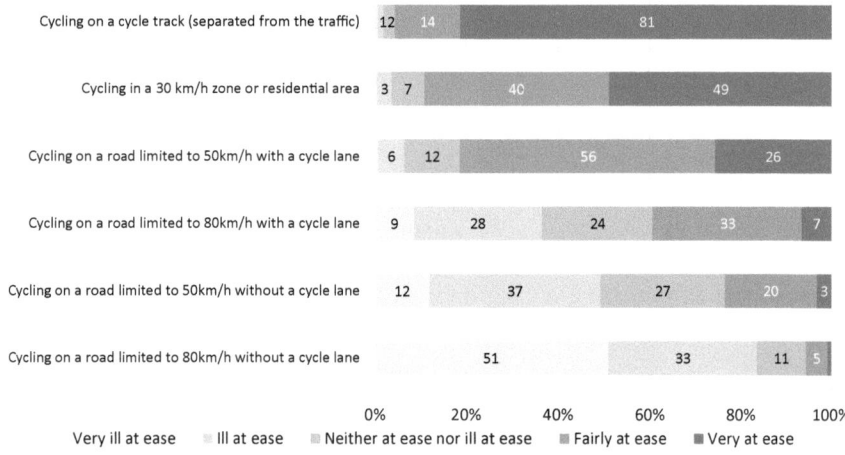

Fig. 7.1 Level of ease according to degree of cohabitation with road traffic

Going to my place of work, sharing a cycle track with children, roller-bladers, walkers, pets, etc. is a disaster. There is a high probability of having an accident. I much prefer to be well integrated in traffic on a wide road, or one with a cycle lane alongside it.

Riding in a 30 km/h zone, or in a residential area, is also perceived positively by the vast majority (89%), followed by two situations, which are characterised by the presence of a cycle lane, i.e. a mark on the ground showing the space allocated to cyclists.[1] While cycle lanes legitimise the presence of cyclists, they do not offer physical protection:

A cycle lane painted on the road unfortunately does not offer enough safety. If you want to motivate people to cycle, you have to invest in infrastructure and routes which are separate from the road.

It's highly psychological. I don't necessarily feel safer, but I feel more legitimate.[…] I'm there, and if the cars get too close, it's clearly their fault. I'm in a cycle [lane]. When there isn't one […] I have the impression that people might think I have no right to be there […] I prefer it when it's better defined. When it's marked out a bit, I'm completely legitimate on the road.

The lowest values are observed when there is no specific arrangement, which is the case in the majority of situations on the Swiss road network: 23% of the cyclists surveyed feel comfortable on roads without a cycle lane where cars can travel at 50 km/h, and only 6% are comfortable on roads outside built up areas (80 km/h), although some highlight the possibility of using such routes when they are not busy (at certain times of the day, for example) (see the following sub-chapter).

---

[1]In common parlance (and in the comments in German and French in the questionnaire or during the interviews), the term cycle route is often used interchangeably to refer to both cycle tracks (physically separated from the traffic) and cycle lanes (markings on the ground). An explanation and a photograph were used to specify this difference in the questionnaire.

These results show the importance of the type of infrastructure in ensuring the attractiveness and safety of journeys made by bicycle. One factor is separation from road traffic, with a clear preference expressed for cycle tracks and then cycle lanes. While the latter do not offer physical separation, they do have the advantage, compared with roads, which do not have dedicated cycle routes, of providing a sense of safety and of legitimising the presence of cyclists. As a result, more cyclists claim to be comfortable on an 80 km/h road, which has a cycle lane than a 50 km/h road which does not. A second factor relates to the speed limit for road traffic. There is a clear hierarchy between roads where the speed is limited to 30, 50 and 80 km/h. The reduction in level of ease at higher speeds is explained by the greater braking distance for motorists, cars overtaking too closely, and the greater difficulty experienced by slower vehicles when entering traffic (lane preselection, etc.)

Other skills were addressed in the questionnaire (Fig. 7.2). Another situation in which cyclists must coexist with road traffic appears to be problematic: roundabouts. More than a third of cyclists say they do not feel comfortable tackling roundabouts. However, a large majority (79%) have no problem scaling an incline that requires physical effort. This can be explained, in particular, by the fact that many participants are regular cyclists, some are motivated by the sporting challenge of *bike to work*, and about one-fifth of them use an electrically assisted bicycle.

Technical knowledge appears not to be very common: a third of the cyclists state that they are not comfortable making small repairs following a puncture, for example. In organisational terms, estimating the duration of a new cycling route is among the skills possessed by the majority (62%). The result is slightly less positive with regards to participants' confidence in their ability to change routes in the event of unforeseen circumstances (46%). This could be explained by the fact that some cyclists have only a limited number of non-busy roads and/or protected routes available to them.

The proportions of e-bikers and conventional cyclists feeling at ease are closely related according to a linear regression ($R^2 = 0.978$). The extent to which people

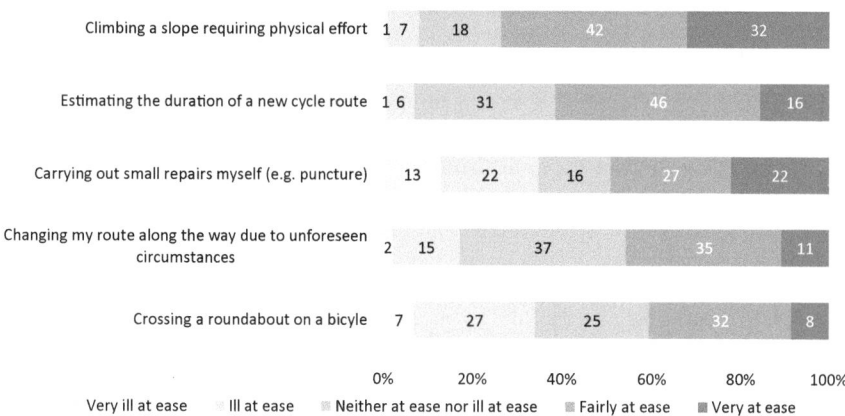

Fig. 7.2 Level of ease in different situations

feel at ease is generally lower among e-bike users, although differences are small. Crossing a roundabout is no easier for e-bike users (38.5 vs. 41.1% for conventional cyclists), although they can accelerate more quickly. The difference between the two groups exceeds 10 points for physical conditions, indicating that electric assistance is not enough to make e-bike users as much at ease as the others (65.0 vs. 75.3%) but also that it helps people with a lower physical condition to cycle. A similar difference is observed with regards to the ability to perform minor repairs (38.5 vs. 51.2%, because e-bikes are more technologically complex or less 'convivial' [3].

7.2   Strategies for Cohabitation with Road Traffic

An additional challenge for utility cyclists is frequent cohabitation with motorised traffic due to a lack of separated cycling infrastructure. Cyclists adopt various tactics [1] in order to cope with the dominance of motorised traffic. On the basis of the 30 interviews carried out at the University of Lausanne [4], a gradient could be identified from the cautious cyclist to the self-confident cyclist. These two ideal types refer to different strategies in motorised traffic—keeping out of the way, or forging a way—and trade-offs between safety and speed.

Cautious cyclists tend to keep out of the way in traffic, to keep a low profile, to take up as little space as possible, and to ride as near as possible to the right side of the road in order to avoid disturbing other road users:

> If the routes I have to use aren't designed for bikes, it's true that I quickly get frightened […]. I try to be as far to the right as possible, to not be in the traffic. So, I don't really take my place as a vehicle. Either I'm on the pavement because I feel really protected there, or really on the edge.

Their feeling of safety in traffic is low, and this influences their choice of route. For them, safety takes precedence over speed and directness. They seek to avoid motorised traffic as much as possible, and are quick to make detours. For this category, cycling infrastructure and separation from cars are crucial. It is worth noting that this behaviour refers to subjective, but not objective, safety. What is seen as a cautious way of riding may actually lead to dangerous situations (e.g. riding too close to the edge of the road makes the cyclists vulnerable to potholes, car-dooring, etc.).

As level of ease increases, cyclists tend not to necessarily keep away from traffic but to adapt their behaviour according to the traffic conditions. In their choice of route, speed becomes more important, and they tend to find safe routes without too much of a detour:

> My commuting trip is along an 80 km/h road with no cycle lane. In the morning, it's not a problem because there's not much traffic. But, in the evening, I always make a detour.

> It depends a lot on the state of the traffic. The cycle routes are not very protected, and there are lots on major routes […] In the morning, it's OK because the motorists are calm. In the evening, they're more excitable… Frankly, you don't feel safe when there are people accelerating.

Confident cyclists are less afraid of the traffic. They tend to be younger, fitter, and more experienced. They do not hesitate to take their place and to claim their legitimacy as road users. They are confident in their bike, in their riding, in their ability to anticipate the rhythm of the traffic. These cyclists use their bike more intensively (during rush hours, bad weather, etc.). Most important in their choice of route are speed and efficiency. For them, a route should be, as far as possible, linear, and without detour, as they feel more capable of facing various situations by forging a way through the traffic:

> *Personally, I think it's fine.[…] I don't mind being a cyclist in the car lane. If I need to take my place, I'll take it. I'm not shy!*

> *At one of the most dangerous places on the journey, there is the metro on one side* [above ground] *and two lanes for cars, one of which crosses the metro tracks. In addition, it's on a descent, where it goes very quickly. Many cars don't pay attention… There is a cycle lane, but it gets cut off in both directions. So at that point, I move into the middle of the road and don't use the cycle lane. It's much too dangerous.*

The level of ease is linked to experience and regularity of cycling, but also depends on other factors (e.g. gender, age, personality, physical condition, equipment). It is worth noting, however, that skills and context are closely linked. Suitable infrastructure is likely to make cycling accessible and attractive to a large part of the population independently of their skills and level of ease.

7.3 Learning by Experience

Travelling by bicycle requires skills or tactics to cope with the traffic conditions and with the level of infrastructure in Switzerland. For some cyclists, participating in *bike to work* represents the opportunity to develop their cycling mobility potential. As discussed previously, some participants refurbished their bicycles or bought one to meet their needs. Others equipped themselves for, or became accustomed to, cycling in bad weather; still others took advantage of this opportunity to extend their journeys.

The event also allows novice or occasional users to acquire certain skills and to become familiar with cycling. In order to assess this, the results presented in this section only relate to participants who stated that they did not travel into work by bicycle before participating in the 2016 event.

In general, participation had a positive impact. As indicated in Chap. 5.1, two-thirds of novices claim, three months after the end of the *bike to work* scheme, that they are now cycling more. This result is explained by several favourable effects of the exercise on how their journey to work is perceived (Table 7.1). 44.5% believe that the journey requires less effort than expected, 35.5% that it is more pleasant and 31.8% that it takes less time:

> *Before* bike to work, *I always thought that my workplace was too far from my home.*

> Bike to work *changed my opinion on using a bicycle and helped me discover a new healthy habit.*

Table 7.1 Impact of participation in *bike to work* on how the commute is perceived among people who did not previously cycle to work

	Less than before (%)	Same as before (%)	More than before (%)
The journey seems to require physical effort…	44.5	42.2	13.3
I enjoy the journey…	4.0	60.5	35.5
The journey seems safe…	11.1	81.2	7.7
The journey seems to take a long time…	31.8	56.0	12.2
I feel at ease in traffic…	8.4	76.1	15.5

The more you cycle, the fitter you are, and the less tiring it is!

I'm just fitter and faster, and so it's less tiring and I find the journey shorter.

My pleasure at being in nature has increased, and I notice the seasons more.

15.5% of novice cyclists say that they feel more at ease, and 7.7% find the route safer:

I'm using the bike more, I feel safer and I'm also much more motivated. I always felt like it was all so hard—but that has changed.

There was a time when I was afraid to ride a bike. That changed with this team event in the sense that, at the start, we often rode together, and then we got more confident with practice. I still feel a lot more at ease on cycle paths through the countryside than I do on the road, but it has improved a lot and I am no longer afraid. But I still get rushes of adrenaline!;-).

However, the feeling of safety is the only one for which there are more negative (11.1%) than positive responses (Table 7.1). Experience varies between participants; negative experiences may lead to a greater awareness of the dangers:

The more I travel around Geneva, the less confidence I have! Risk taking is proportional to participation!!!

A rush of confidence when you're in traffic is not necessarily a good thing, because you take more risks! And motorists are really aggressive with cyclists, I have been insulted several times!

I became increasingly aware of danger spots, and had a collision with a car/distracted driver at a roundabout.

Experience facilitates the acquisition of knowledge of the terrain and potentially problematic situations:

My awareness of traffic 'black spots' has increased so much that I could map them out!

I realised how many times my rights are violated, for example non-respect of right of way, etc.

It also helps cyclists to understand the point of view of the various road users:

I now understand traffic problems from a cyclist's perspective, but also from that of the motorist, and try to behave in the best way from both perspectives.

In general, I have become more aware of other road users.

Through our participation in bike to work, *we have become more aware of the dangers that exist on the way to work. We drive our cars more respectfully with regard to cyclists.*

The lower scores for safety questions are explained by the fact that they depend not only on the skills and level of ease of the participants but also, and especially, on the existing infrastructures and facilities, and on the behaviour of motorists. Here we come up against the limits of promotional campaigns, which focus exclusively on individual mobility potential and not on a territory's hosting potential:

I see the biggest problem for cyclists in Switzerland as being poor infrastructure. There are hardly any cycle tracks separated from the road in the Basel and Zurich region. A cycle lane painted on the road unfortunately does not offer enough safety. If they want to motivate people to ride a bicycle, they should invest in this type of infrastructure; bike to work on its own is unfortunately not enough.

Bike to work *was a highly motivating factor for cycling. Unfortunately, the conditions are not really conducive to it. There should be cycle tracks separated from the road for real safety, which would require extensive reorganisation but would help increase the proportion of bicycles. Some cities have started doing this, but sadly not Lausanne and the wider region.*

It's great to be made more aware of cycling during the spring. However, the safety levels on my way to work have not improved. This is why I didn't keep using my bike to get to work after the bike to work campaign. But in my free time, I like to look for nice routes in the countryside and on quieter and less crowded routes.

The interviews and comments revealed the crucial importance of the choice of route. A low level of ease encourages cyclists to choose a safe route, away from motor traffic. However, research carried out in Geneva has shown that travel habits are insufficiently taken into account in the promotion of cycling [2]. People who new to cycling tend to use the routes they used to travel by car or public transport, although these may not be the most suitable.

Choosing a route sometimes requires an investment of time. It is done by trial and error, by consulting paper or online maps, by testing out the different options, by moving away from the most frequented routes, and by avoiding certain places during rush hour:

Thanks to bike to work, *I discovered a safer route to get from my home to work. It involves paths that are inaccessible to cars, which are parallel to the road, and I didn't know they existed.*

Bike to work *allowed me to 'visit' my city via different routes. And to find the least busy ones...*

With practice, you can optimise your journey to save time and/or increase safety and/or enjoyment. Personally, I have found several variants for my journey from home to work.

This learning can take on a collective dimension, with novice cyclists learning from more experienced colleagues. The following discussion, which took place in March 2018 on a social network between cyclists from the University of Lausanne (UNIL), summarises the skills required and the value of sharing experiences with other cyclists:

Person 1: *Would it be possible to do 'accompanied' or 'guided' journeys from UNIL? I would like to do my journeys by bicycle, but I don't know how to behave (with a two-wheeler) on the road. The big roundabout at […] is the main reason why I come by bus and not by bicycle. In any case, I'm not sure it's a 'good route' for a cyclist…*

Person 2: *There may be cycle tracks and underpasses that would enable you to avoid cycling via the […] roundabout.*

Person 3: *I don't know exactly where you come in from but one possibility, […] is to take the […] cycle paths. It does take you on a detour, but there's hardly anyone on this route. On the other hand, if you want to go to […], you have to weave along UNIL's cycle paths. I don't know if that is of help to you, if not, let me know!* [A section of a map illustrating the route is included here].

Person 1: *It's already clearer;-).*[…] *There are lots of us who 'don't know how to ride a bicycle on the road'. It's a skill that can't just be taken for granted, even if you have a driving licence. I think that offering to accompany people could be a great way of promoting it […] and would encourage cycling […].*

Person 4: *Totally! We'll have to put it in place. If you cycle regularly, there are a few 'keys' to understanding motorists and putting yourself in as little danger as possible. But the first thing is to make sure you can be seen (lights, etc.), to pay attention (watch cars carefully and anticipate their movements) and, above all, signal your own changes of direction, etc. And it's also true that the more your journey involves car-free routes, the lower the risk;-) In any case, let us know your route, so that a regular cyclist […] can go with you and show you the best paths.*

In June 2018, the Swiss Federal Institute of Technology Lausanne, which is located on the same campus as the University of Lausanne, offered its students and employees three types of route: 'ride and discovery' (traffic-free route, in a more user-friendly and secure setting); 'fast and peaceful' (route combining speed and safety) and 'traffic and direct' (fast route using the major roads within the agglomeration). The online route maps are supplemented by the opportunity to take part in accompanied rides with a specialist coach. The initiative is interesting in the sense that it provides cyclists with a better understanding of the territory they are travelling across and offer them a variety of routes according to their level of ease.

$$* \quad * \quad *$$

Cycling requires skills that go far beyond just knowing how to ride a bicycle and having certain physical abilities. In general, there is a great sensitivity to infrastructure due to the vulnerability of cyclists in motor traffic. The cyclists' levels of ease are highest in the case of cycle tracks, which are separated from road traffic. It then decreases according to the degree of coexistence with traffic as well as the speed and volume thereof.

Depending on their level of ease, cyclists adopt various strategies when choosing their routes. Each figure represents a trade-off between safety and speed. Cautious cyclists tend to fade into the background in traffic. Strategic cyclists adapt their behaviour according to the traffic conditions. Confident cyclists are less afraid of sharing the road with traffic, and do not hesitate to take up space and to assert their legitimacy as a fully fledged road user.

There is thus a gradation in the levels of ease and expertise between novice cyclists and experienced cyclists, with the former being much more sensitive to traffic conditions. For novice cyclists, *bike to work* is an opportunity to try out cycling as a means of transport. When the perception of the journey changes, the change is generally positive in terms of duration, pleasantness or effort required. The effect is much less positive in terms of safety, which points to shortcomings in cycling infrastructures and facilities in Switzerland.

References

1. M. de Certeau, *The Practice of Everyday Life* (University of California Press, Oakland, 2013)
2. M. Flamm, *Etude sur les choix d'itinéraires des cyclistes à Genève* (Etat de Genève & MICODA, Genève, 2014)
3. I. Illich, *Tools for Conviviality* (Calder and Boyars, London, 1973)
4. A. Schmassmann, *Vers un environnement cyclable de qualité: un diagnostic du campus de l'Université de Lausanne* (Institut de géographie et durabilité, Lausanne, 2018)

Chapter 8
Motivations

What motivates a commuter to get on their bike to travel to their place of work? The literature highlights not only utilitarian elements but also the meanings and sensory experience of cycling. What about within our sample? What are in addition the motivations to take part in *bike to work*?

First, we propose a quantitative analysis of the significance of the various motivations. Each major dimension is then more finely interpreted using the interviews and comments left in the questionnaire. A typology in four categories is established according to the motivations of the *bike to work* participants. The reasons for taking part in this action are presented at the end of the chapter.

8.1 A Plurality of Rationales for Cycling

The appropriation of a mode of transport relates to elements that are perceived positively—which we approach here as motivations—or negatively—i.e. barriers, which are the subject of the next chapter.

The respondents were asked to judge whether different factors influence their decision to cycle to work (Fig. 8.1). Almost all of them (98%) state that being able to exercise is a motivation. Other criteria are favoured by more than four out of five participants. These are flexibility and freedom (90%), the pleasure linked to the journey (feeling of being in the fresh air, landscape) (88%), respect for the environment (88%), and the ability to take their mind off things and disconnect from work (80%).

Saving time and money are also valued by a majority of respondents, although these preferences are less marked than for the previous criteria (60% and 53%, respectively). This can be explained, in particular, by the fact that some cyclists travel long distances during part of the year and use other modes the rest of the

P. Rérat, *Cycling to Work*,
SpringerBriefs in Applied Sciences and Technology,
https://doi.org/10.1007/978-3-030-62256-5_8

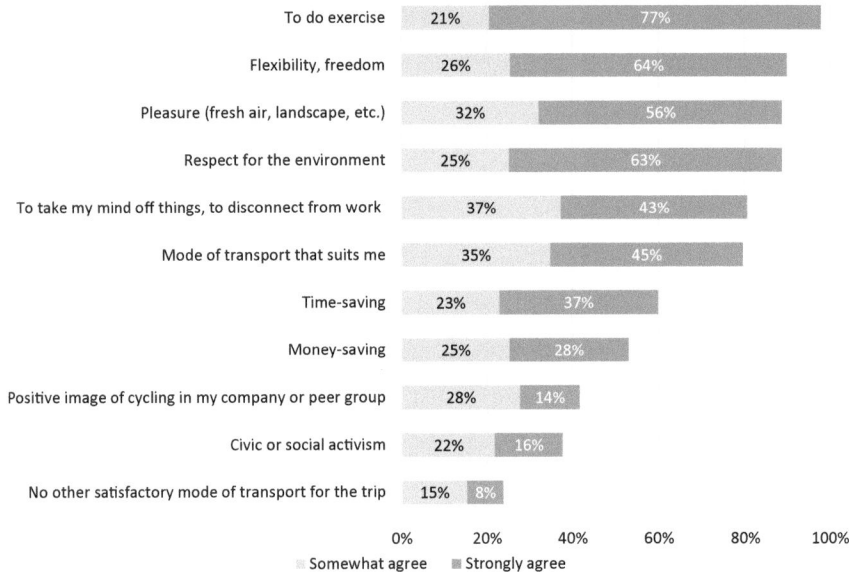

Fig. 8.1 Motivations for using a bicycle for the home–work commute

time. A difference is also noted depending on professional status: students give more weight to these two criteria (72 and 59%).

Conversely, less than half of the respondents mention social activism as a reason for cycling, or consider it important that cycling has a positive image among their friends or colleagues. Less than one in four considers a lack of adequate transport on the route to be important, which suggests that cycling is a choice (rather than a necessity) for the vast majority.

Motivations stated by e-bike users and mechanical cyclists are highly correlated (linear regression; $R^2 = 0.982$). Differences are below one percentage point for five items: exercise, flexibility and freedom, environment, saving money, and the image of the bike in the social and professional circle. There are only four variables where the difference exceeds two points: E-bike users are more likely to mention the pleasure of riding (92.6% vs. 88.0%) and the opportunity to take one's mind off things and to relax (85.0% vs. 79.7%), but less likely to answer time-saving (53.5% vs. 61.1%) or social activism (35.5% vs. 38%). These differences do not seem related to the electric assistance but are explained by the fact that more e-bikers have access to a car and that they take it as a point of comparison. This explains why the experience of riding is more important for them, but they are less likely to see it as time saving, and also why they cycle to do exercise and in respect for the environment in the same proportion as conventional cyclists. The absence of difference in the importance given to exercise indicates that the emergence of electrically assisted bicycles allows certain people to continue to travel on two wheels and to engage in physical activity to keep themselves in shape.

Some of these variables are similar to each other and measure similarly to the rationales for cycling to travel to work. It is possible, for increased readability, to summarise the motivations across three axes: well-being, civic engagement and independence.[1] These three components are quite similar to Cresswell's conceptualisation [1, 2] (see Sect. 2.3.3), which are, in order, the experience of the journey (well-being), the meanings represented by the choice of mode of transport (civic engagement), and the movement itself and its factual characteristics (independence).

8.2 Well-Being

The first component of the analysis relates to the elements, which contribute to an individual's well-being: the opportunity to exercise, the pleasure of riding a bicycle, and the opportunity to take their mind off things and to disconnect from work. It relates to the experience of mobility and, more specifically, to the benefits of cycling, both physical (exercise) and psychological (pleasure, the feeling of being in the fresh air, means of escape).

The concept of well-being covers hugely variable practices and ambitions. This is the case with exercise, which emerges here as the most important motivation and which frequently appears in the comments (for example, *'If you rest, you rust!'*). It takes very diverse forms, from brief and moderate activity (including for people suffering from health problems) to intense sports training:

> *Since I tore my cruciate ligament, I can no longer run. Riding a bicycle, even an e-bike, allows me to do some sport.*

> *As a Diabetic, cycling has a positive effect on my blood sugar. I try to cycle as often as possible.*

> *It's been eight years now that I've been participating in* bike to work, *but it's only been three years since I started using my bike regularly to get to work. What helped motivate me a lot was an optional health test at work [...]. I was in the orange for a little while three years ago. I think now everything should be in the green again.*

> *I often combine my journey home with a long exercise session (Instead of 8 km, I do 50–60 km). So I can do endurance training while it's still daytime.*

> *I do long-distance triathlons. The route to work is part of the training. some glorious moments in the morning at daybreak!*

---

[1] A principal component analysis (PCA) was conducted with SPSS on the 11 motivations with orthogonal rotation (varimax). An initial analysis was run to obtain eigenvalues for each component in the data; three components had eigenvalues over Kaiser's criterion of 1 (2.69, 1.41 and 1.26 respectively). They explain 65% of the variance (33.5%, 17.6% and 14.1%) and their Cronbach's Alpha ranges from .65 to .68, which can realistically be expected when dealing with psychological constructs and when there is a small number of items on the scale [3]. The variables 'money-saving' and 'no other satisfactory transportation means' are excluded as they only add a weak explanatory contribution (cut-off points for factor loading >0.5). The variable 'transportation mode that suits me' is set aside as there is no significant link with any of the three axes.

Many participants noted that available time throughout the day is a scarce resource. Cycling to work allows people to combine exercise with compulsory travel and to integrate physical activity into the daily routine of a population which is primarily engaged in sedentary and office jobs. It is a *'trick'* that saves time (although not travel time in the strictest sense) by avoiding having to devote an additional slot to (travelling to) sporting activities. This optimisation of their schedule allows them to spend more time at home with their family, for example:

> *I combine commuting to work and sport, so that I don't use my lunch break or my evening to do sports. I have more time to spend with my family.*

> *It's hard to find time to do sport these days. Using my compulsory travel to exercise by cycling represents a significant gain in my schedule even if I have to spend more time to make the trip in comparison to the train or car.*

> *It avoids having to spend more time doing sport after driving home from work… ultimately, I have saved time for other things.*

> *Cycling is a wonderful way to clear your head after a stressful day at work, sitting at a desk. It is also good for physical fitness and goes almost unnoticed with regard to family life.*

This last quote introduces another aspect of well-being, which is as much physical as psychological. It is about the pleasure of cycling, feeling the body in motion, and experiencing the environment and countryside through the different senses:

> *Cycling is simply fun!*

> *VÉLOVE!* [vélo meaning bike in French]

> *I regularly make detours of more than an hour on my way to work just for fun.*

> *Being on the move in the fresh air in the morning is the first highlight of the day!*

> *Cycling to work […] is a great way to discover and gain awareness of the environment. We are in contact with the world around us, we experience the cold and the heat, the smells, the sounds, the songs of the birds. It's good for balance. Work-life balance is improved because cycling goes hand-in-hand with physical effort and pleasure.*

Finally, time spent travelling by bike has added value because it allows users to get away from everyday life and to take time for themselves. It can represent a bit of breathing space before or after the working day, a decompression chamber allowing them to shake off their worries and problems:

> *I feel like I start my working day in a better mood and with more energy when I cycle in.*

> *I love my bike! to go to work by bike is an enjoyable moment in my day and makes me (almost) fully disconnect from work.*

> *When I cycle home, I can work my way through a whole series of problems while pedalling so that I don't bring them home with me. I arrive home happier.*

For some, the break offered by the cycle journey also represents a way of avoiding journeys by public transport or by car, and of disconnecting by focusing on their own physical movement:

> *I avoid the depressed looks of people on the bus, and what's more, as I pass by the lake, I can start the day well with oxygen in my brain.*

I couldn't use the car or the bus every day [...], I would get depressed. Cycling is synonymous with freedom and physical well-being, but above all it allows me to forget about my worries and negativity.

Cycling, whether using an e-bike or traditional bicycle, gets you on the move, and so you can focus on what you are doing without having to look at your phone and without having the feeling of waiting around at the station, bus stop, or in traffic jams.

Overall, the experience of cycling appears to be extremely beneficial for well-being. First of all, commuting by bicycle provides an opportunity of—sometimes incidentally—doing exercise. This can be the primary goal but also, and above all, a benefit that accompanies the use of the bicycle. It is a way to transform a physical activity into an enjoyable activity, to squeeze it into a busy day and to create the conditions for it to become a habit rather than a chore or a constraint [7]. The well-being derived from cycling also has a psychological dimension. Far from being seen as a waste of time, cycling is a pleasurable practice, a positive sensory experience of one's surroundings, and an opportunity to get away from everyday life, to disconnect or to find time for oneself. The valorisation of travel time has been a focus of research on the use of public transport in particular. It allows the user to engage in various activities (reading, working, resting, social media, etc.). The above results show that cycling trips are valued from the point of view of both physical and mental well-being. In addition, from this perspective, the boundaries between cycling as a sport, leisure activity and means of transport become more fluid and blurred.

8.3 Civic Engagement

The second component of the analysis is civic engagement, which relates to respect for the environment, social activism and the image of cycling. Cycling represents an ecological and sustainable means of transport (*'clean'*, *'CO_2free'*), both locally and on the global scale. It is also a way of reclaiming quality of life – both now and in the future—in cities:

I hope that bicycles will be used more in the future and that we will be able to breathe better, for ourselves but especially for our children.

In my opinion, in urban traffic and in densely populated areas, cycling is one of the best ways to overcome transport problems [...]. Both individual motorised traffic and public transport are currently reaching maximum capacity in many places, and there is no more space to increase this capacity in a substantial and sustainable way.

Although I'm not at all a sporty person, I love cycling. I like to do something for the environment, for the quality of life in the city, for our health [...], and I like the experience of being out in the open air, and the feeling of working as a team. Unfortunately, I understand that I put my life at stake every day by going out cycling, despite the helmet, bright clothes and lights.

In this case, the practice of cycling represents a way of *'embodying citizenship'* [5] and is seen as a responsible act. It is also seen as a way of setting an example to colleagues or children and to demonstrate that alternative forms of mobility to the car are possible and desirable:

> *My route to work is much shorter by bicycle [...] and much more pleasant [...]. And I think it essential to set an example, to show that it is entirely possible to travel by bicycle in Lausanne, even for getting to business meetings.*

> *An additional motivation is to be a role model for my children. I would like them to see cycling as an obvious alternative for getting to school, for getting to work.*

> *I want to be an example to the people around me and be able to say: it works just as well this way!*

Cycling can be perceived as being more compatible with a human's '*biological rhythm*'. This echoes Illich, for whom travelling using muscular energy is a way to reconnect with the environment (Illich 1973) and with natural elements such as weather conditions:

> *To ride a bike allows me to rediscover a speed of life adapted to a human biological rhythm... Completely to the contrary of the car, which transforms the human being into a stressed person.*

> *The increasing mobilisation through cars, motorbikes and lorries bothers me greatly. Against this I would like to set an example in terms of how I respect the environment and say: it is easy to do like this. To go everywhere powered by our muscles, so we are free, independent and stay fit. To be outside, to feel the sun and the rain on your skin, to let the wind whistle in your ears, to find your way in the fog. This is life!*

Environmental issues, however, are often a secondary motivation that overlaps with others when choosing to cycle:

> *There are also vaguely environmental reasons. They are not necessarily primary reasons. but it's a pleasure to tell myself that all of my trips are pretty much carbon neutral.*

This second component, civic engagement, explicitly relates to extrinsic motivations,[2] as civic engagement involves the pursuit of external goals. This is illustrated by expressions such as '*for our children*', '*for the environment*' or '*for the quality of life in the city*' in the above quotes. The adoption of a sustainable means of transport is reinforced by the desire to demonstrate that cycling is a credible and attractive alternative.

8.4 Independence

The third component relates to the practical elements of cycling as an individual means of transport: it saves time and is characterised by freedom and flexibility. Cycling is thus seen as '*simple*', '*flexible*', '*fast*', facilitating '*door-to-door journeys*', and free of timetables:

---

[2]Motivations can be described as extrinsic or intrinsic depending on the reasons and objectives underlying an action. An extrinsic motivation is defined as a motivation to do something based on an external constraint or objective, while an intrinsic motivation relates to the interest and pleasure that an individual finds in an action [6].

Simple and flexible, you can quickly and easily stop at any store, without looking for a parking space and without having to pay.

It is possible to park anywhere, to quickly do some shopping, or something else along the way...

For some participants, cycling is not a political act and is a normal and commonplace activity (while for others, as we have shown, this practice has a political significance):

Cycling is simple and should stay that way. I don't Do it because of ecological or political conviction, or for any other 'deep' reasons.

Cycling is characterised by advantages that are often compared against the difficulties associated with other means of transport. It overcomes the constraints of both the car (traffic jams, search for a parking space, cost, etc.) and public transport (saturation, inflexibility of timetables and routes, etc.):

Cars get stuck in traffic all the time. when I cycle, I arrive everywhere on time!

Flexibility, being able to leave at any time, whereas with the metro or the bus, you have to leave your home at a specific time, at a specific minute. If you miss it, you have to wait [...], it quickly gets complicated. While on a bike, you can leave whenever you want. Anyway, I never arrive at work at the same time!

It's simpler. You're not dependent on public transport, you're free... You're alone on your bike, you're not in a crowded metro full of people.

I cycle out of laziness. All other modes of transport are too restrictive. Car: searching for a parking space, traffic jams, rush hour. Public transport: stick to the timetable (or miss the bus), you have to change bus/train, no overnight service. Cycling: door-to-door and round-the-clock.

When compared with walking, it is cycling's greater efficiency—in terms of speed and effort—that is highlighted:

I live more or less two kilometres from the university. On foot, it's still a little over twenty minutes. By bike, it only takes five to seven minutes!

Independence is a characteristic of individual modes of transport as opposed to public transport. Like cars and motorised two-wheelers, bicycles allow users to choose the place of origin and destination of a journey, and to be independent in their choice of route and timetable. In urban areas, the benefits of cycling are reinforced by comparison with to motorised transport, due to the short distance of journeys and to barriers to car traffic (congestion, parking, etc.).

8.5 A Typology of Cyclists According to Their Motivations

As the above quotes reveal, people travelling to work by bicycle are not a homogeneous group, and multiple meanings are assigned to the practice. A typology allows us to lift the veil on this diversity by identifying four categories of cyclists according

Table 8.1 Typology of cyclists according to their motivations

	% of the sample	Well-being	Civic engagement	Independence
Active cyclists	29	+	0	–
Civic cyclists	17	–	+	0
Individualistic cyclists	14	+	–	+
Enthusiast cyclists	40	+	+	+ +

Note "+": More than average; "-": Less than average; "0": Close to average

to the importance given to the three axes of well-being, civic engagement and independence[3] (Table 8.1). These categories represent various constellations in which movement, meanings and experiences are combined. They have then been crossed with socio-demographic characteristics, equipment and mobility practices in order to better understand their specificities (Table 8.2). According to statistical tests, the differences are significant; this is partly due to the large size of the sample. In the comments, only the most important differences are mentioned.

The first category brings together active cyclists, who make up 29% of the sample. They are more interested than average in aspects of cycling related to well-being, such as physical activity (from moderate to athletic) and the pleasure of riding a bicycle. The characteristics associated with movement in the strictest sense—travel time and flexibility—are less important to them than for the other groups.

There is a trend in this category towards an over-representation of men, participants living in households with children, people aged over 40, those with a professional education, and residents of suburban, peri-urban and rural communities. Their journeys are significantly longer than the average. Outside of *bike to work*, they opt more regularly to use cars, and their cycling is more recreational than utilitarian. Participating in *bike to work* is an opportunity to practice cycling in order to stay in shape or to get away from everyday life. Their long journeys mean this category demonstrates the greatest seasonality of practice.

Civic cyclists (17%) make up the second group. They are more likely than other groups to mention extrinsic motivations, such as environmental concerns. The practice of cycling thus takes on a more political meaning. The importance they place on independence is similar to the sample as a whole. However, this category is the only one to be below average for concerns related to well-being.

This type tends to have an over-representation of men, city dwellers, young people and people with a high level of education. Cycling is more utilitarian and less of a hobby in and of itself (leisure rides or sport). They have less access to cars than the average and do the shortest commuting journeys. Their participation in *bike to work* is based—in greater proportions than for the other groups—on the opportunity to reassert the importance of cycling and on responding to invitations from colleagues.

---

[3]On the basis of the z-standardised factor loadings obtained by each participant for the three factors, a hierarchical cluster analysis (Ward logarithm) was conducted in order to establish a typology. The number of groups (4) was chosen by examining the agglomeration schedule. A demarcation point was observed between 4 and 5 groups.

Table 8.2 Main characteristics of the four types of cyclists defined according to their motivations

		Active cyclists (%)	Civic cyclists (%)	Individualistic cyclists (%)	Enthusiast cyclists (%)	Total (%)
Profile						
Gender	Women	35.6	35.9	42.2	48.9	41.9
	Men	64.4	64.1	57.8	51.1	58.1
Age	15–24	4.2	4.6	4.3	3.2	3.9
	25–39	30.2	39.6	41.8	36.4	35.9
	40–54	49.9	41.8	42.7	45.8	45.9
	55 and above	15.8	14.0	11.2	14.6	14.4
Level of education	Low	21.3	17.1	14.4	17.2	18.0
	Average	32.7	24.2	27.3	26.8	28.2
	High	46.0	58.6	58.3	56.0	53.8
Residential context	Urban	22.5	51.9	39.6	46.4	39.5
	Suburban	49.1	38.4	48.5	44.0	45.2
	Rural	28.4	9.7	11.9	9.6	15.3
Household	No children	45.8	53.7	53.6	50.9	50.3
	With child(ren)	54.2	46.3	46.4	49.1	49.7
Journeys by bicycle						
Seasonality	Only during summer months	44.1	18.0	22.9	10.9	23.4
	All year	55.9	82.0	77.1	89.1	76.6
Frequency of cycling on commuting journeys	Occasionally	42.4	14.9	16.1	7.0	19.9
	Regularly or most of the time	57.6	85.1	83.9	93.0	80.1
Distance of commuting journeys	Duration of return journey	65'	33'	46'	40'	47'
Access to other means of transport						
Access to car	Yes, always	63.9	46.5	50.1	45.7	51.8
	No or on demand	36.1	53.5	49.9	54.3	48.2
Annual public transport pass	Yes	31.3	38.8	38.3	32.7	34.1
	No	68.7	61.2	61.7	67.3	65.9

Note The statistical tests (chi-squared except for distance, for which the ANOVA method was applied) show that the differences are significant for all variables ($p < 0.001$)

The third category, individualistic cyclists (14%), emphasises the personal benefits of cycling, both in terms of pleasure and independence. The weighting they give to civic engagement is much lower than for the other three groups.

The under-40's and people with a high level of education are somewhat over-represented in this third type of cyclist, although in other ways, the group is close to the values of the sample as a whole. Cycling appears to be a practical and efficient way of getting around, but is less associated with sporting or leisure activities. They are less likely to see their participation in *bike to work* as an environmental action.

Finally, enthusiast cyclists make up the largest group (40%). They differ from the other categories by assigning more weight than average to all three components. The independence provided by cycling, civic engagement and personal well-being constitute the three pillars of a practice which is strongly embedded in their daily lives.

There are more women in this category than average, and more city dwellers. They cycle predominantly or even systematically, and make little use of cars. Their use of bicycles is particularly intense for utility purposes (work, shopping, getting to a leisure activity). Their participation in *bike to work* is the most regular and aims, in particular, to reassert the importance of cycling and to motivate their colleagues.

8.6   Participation in Bike to Work

The respondents commented on the factors that motivated them to take part in *bike to work* (Fig. 8.2). The first factor is consistent with the results presented above: the opportunity to do some exercise, which is cited by more than 6 in 10 respondents. The other intrinsic (micro-scale) motivations are the least important of the factors mentioned—these include the possibility of winning a prize, and the opportunity to

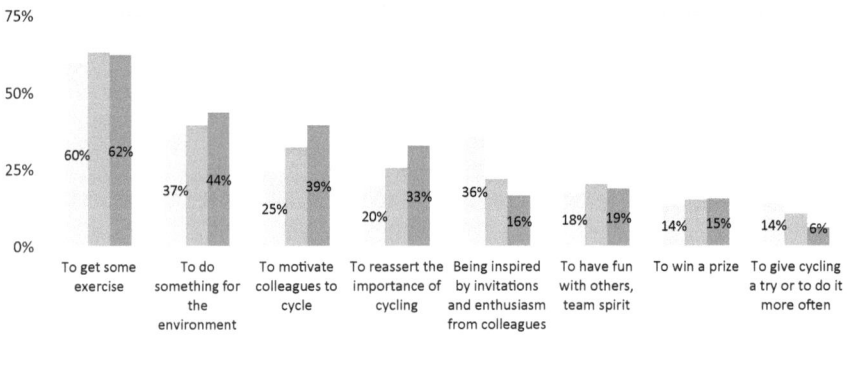

Fig. 8.2 Motivations for taking part in *bike to work* according to number of years of participation (maximum of three responses per participant)

give cycling a try or to do it more often. Two extrinsic (macro) motivations rank second and fourth in importance; these include the chance to do their bit for the environment and to reassert the importance of cycling. As bike to work is a group activity, there are also two (meso) factors linked to being part of a team: encouraging colleagues to cycle or responding to their invitations.

Weighting the motivations according to the number of times the respondents have taken part in the scheme shows relative stability for physical exercise, feeling part of a team and winning prizes. Three factors stand out among the motivations for participating in the initiative at least three times: doing their bit for the environment, motivating colleagues and reasserting the importance of cycling. Trying out cycling and being encouraged by colleagues are, logically, primarily motivating factors for people taking part in the campaign for the first time.

Three effects of *bike to work* can be identified according to the scale on which they are deployed: the trigger effect (micro), the mobilising effect (meso) and the awareness-raising effect (macro).

The trigger effect concerns two types of participants. First, those people who cycle seasonally, and those who are not primarily utility cyclists. According to participants' comments, the campaign encourages them to '*get your bike out*', '*get the season started*' and '*overcome the laziness stopping you from cycling*':

> Bike to work *motivates me every year in the spring to get my bike back in working order. Then it's available to use again and so I use it more, all summer and autumn.*

> Bike to work *gives me the 'kickstart' to mostly cycle until the end of the year.*

For others, the scheme reintroduces cycling after a period of abandonment associated with a significant life event, such as moving house. It acts as a '*spur*', and gives a '*new impetus*' to start cycling again:

> *It motivated me to ride my bike more again. When I lived in town, biking was MY means of transportation. I didn't need a car. In the countryside, this habit changed. Taking part in bike to work gave me a new impetus.*

> Bike to work *was the boost I needed to get on my bike.*

> Bike to work *was the trigger. The idea, the intention to cycle to work had already been there for a longer time.*

This trend is reinforced by the mobilising (meso) effect of *bike to work*. The creation of teams makes it possible to create a '*group dynamic*', to '*share respective experiences*' and to facilitate contact between colleagues around a '*common hobby*'. Some respondents highlight friendly rivalry with their peers as a motivation, resulting from the competitive element of the initiative and the kilometre count:

> *During the* bike to work *month, we challenged ourselves with colleagues and we cycled every day, even in heavy rain, which I don't normally do. With the right equipment, that was no problem, it was fun and I noticed that we weren't the only cycling "fools". We pushed ourselves to the limits by adding additional kilometres to lengthen the route. We were proud! Generally speaking, I just felt so good and happy when I arrived in the morning and looked forward to riding my bike again in the evening.*

> *Participating in* bike to work *somewhat piqued my ambition because we compared who on the team was riding the most.*

However, this effect can sometimes be experienced in a negative way and can take the form of peer pressure:

> *The fact of "having to" come in by bicycle for a whole month was not positive for me. I forced myself to cycle in, especially because of my teammates and my other colleagues. At the end of the "imposed" month I didn't go back to using my bike to travel in to work.*

Finally, *bike to work* has an awareness-raising effect (macro) in the sense that participation is seen as a means of making cycling visible and legitimate, of creating a feeling of belonging to a larger group, *'being part of something'*, *'casting a vote for cycling'*[4]:

> *This gives cycling a more positive impact. people talk about it a bit more at work.*
>
> *I was and still am passionate about cycling. I take part to highlight the importance of cycling as a means of transport.*
>
> *I feel connected with my colleagues who also took part in bike to work. I feel less isolated in road traffic.*
>
> Bike to work *is a way to increase visibility. In addition, perhaps it will motivate new people to do it, but that's less important to me.* [...] It's a way of casting a vote for cycling.

* * *

The participants in *bike to work* put forth three main rationales for bicycle use: the independence it provides as an individual means of transport (freedom and flexibility), civic engagement through adopting sustainable mobility (global and local environmental issues), and well-being (both physical and mental).

This last point is important because it sheds light on the experience of travelling by bicycle. Travelling time is valorised thanks to the benefits that commuters can get from it (exercise, escape, disconnection from work, experience of the natural environment, etc.). This result contrasts with the traditional approach to transportation, which considers mobility to be a derived demand—i.e. a consequence of the desire to reach a certain place—or as a neutral experience that 'costs' time. It also shows that the boundaries can be blurred between utility, sport and leisure cycling.

The commuters' motivations, on an individual basis, intersect with certain political concerns—such as public health and climate change—that are leading an increasing number of societies and organisations to encourage cycling. An additional and important dimension that is worth mentioning is the pleasure linked to the experience of riding a bicycle. To borrow Bjarke Ingels' [4] term, cycling is an example of 'hedonistic sustainability'.

There are diverse meanings associated with cycling. Four main categories of cyclists can be identified according to their motivations: active, civic, individualistic and enthusiast cyclists. Their participation in *bike to work* depends on a range of motivations, largely based on the regularity with which they cycle. It can be seen as the quest for a trigger effect (taking up cycling (again)), a mobilising effect (creating

---

[4]Certain institutions—like the Swiss Federal Institute of Technology Lausanne and the University of Lausanne – organise events such as parades at the end of *bike to work* in the aim of showing the importance of cycling.

a group dynamic) and an awareness-raising effect (demonstrating the importance of cycling). However, the practice of cycling comes up against a number of barriers, which are detailed in the next chapter.

References

1. T. Cresswell, *On the Move: Mobility in the modern Western world* (Routledge, 2006)
2. T. Cresswell, Towards a politics of mobility. Environ. Plan. D Soc. Space **28**(1), 17–31 (2010). https://doi.org/10.1068/d11407
3. A. Field, *Discovering Statistics Using SPSS* (Sage, London, 2009)
4. B. Ingels, Hedonistic sustainability. (2011). https://www.ted.com/talks/bjarke_ingels_hedoni stic_sustainability. Accessed 01 Aug 2020
5. J. McKenna, M. Whatling, Qualitative accounts of urban commuter cycling. Health Educ. **107**(5), 448–462 (2007). https://doi.org/10.1108/09654280710778583
6. R.M. Ryan, E.L. Deci, Intrinsic and extrinsic motivations: classic definitions and new directions. Contemp. Educ. Psychol. **25**(1), 54–67 (2000). https://doi.org/10.1006/ceps.1999.1020
7. P. Walker, *Bike Nation: How Cycling Can Save the World* (Yellow Jersey Press, London, 2017)

Chapter 9
Barriers

What are the barriers commuters encounter in their practice of cycling? To which elements and scales do they refer? How can they be overcome?

As in the section on motivations, this chapter begins with a quantitative overview. The main barriers are then explored in depth using interviews and respondents' comments. They are addressed in terms of the scale at which they operate: that of the individual, their household or, more generally, the context in which cycling takes place.

9.1 Barriers to Cycling

The results relating to motivations are inherently positive because they relate to the elements, which favour the practice of cycling. However, this practice also comes up against barriers, which are quite different to the types of motivations that have been discussed. First, while barriers can be general in scope (like motivations), they can also relate only to specific situations (for example, bad weather or the transportation of objects). Second, the scores observed are lower than for the motivations, which are explained by the fact that the respondent population is made up, to admittedly different degrees, of cycling enthusiasts and has, at least in part, overcome these problems.

It is nonetheless important to consider the barriers to cycling, to identify not only the frictions experienced by people who have chosen to commute by bicycle but also the issues that need to be overcome in order to expand utility cycling to a wider

P. Rérat, *Cycling to Work*,
SpringerBriefs in Applied Sciences and Technology,
https://doi.org/10.1007/978-3-030-62256-5_9

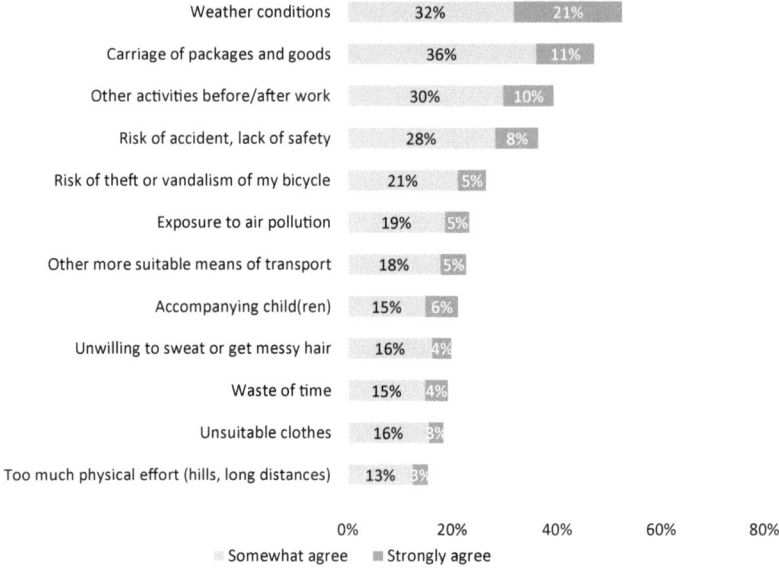

Weather conditions — 32% | 21%
Carriage of packages and goods — 36% | 11%
Other activities before/after work — 30% | 10%
Risk of accident, lack of safety — 28% | 8%
Risk of theft or vandalism of my bicycle — 21% | 5%
Exposure to air pollution — 19% | 5%
Other more suitable means of transport — 18% | 5%
Accompanying child(ren) — 15% | 6%
Unwilling to sweat or get messy hair — 16% | 4%
Waste of time — 15% | 4%
Unsuitable clothes — 16% | 3%
Too much physical effort (hills, long distances) — 13% | 3%

0%   20%   40%   60%   80%

■ Somewhat agree ■ Strongly agree

Fig. 9.1 Barriers to using a bicycle for the home–work commute

population. Indeed, potential cyclists—who are less convinced or less experienced than the current users—are more likely to consider these barriers as prohibitive [2].[1]

There is no correlation between the typology of the cyclists according to their motivations presented above and the importance to cyclists of the barriers. The factor that has the greatest explanatory power regarding barriers is the frequency of cycling. The higher the frequency, the less significant the barriers appear to be. This result is in line with an observation made by Flamm (1, 231) that 'the more we master a mode of travel, the more we are able to appreciate it, and the more we appreciate a means of transport, the more likely we are to use it and thus to improve the skills that enable us to use it'.

The factors hindering cycling can be summarised into four groups (Fig. 9.1): weather conditions, logistical constraints, safety (of the individual and their vehicle) and comfort.[2] Certain aspects have already been raised in the discussions around

---

[1]In this sample, barriers and motivations are not correlated. The most explanatory factor regarding barriers is the frequency of use of bicycle commuting before the *bike to work* campaign. The higher the frequency, the lower the importance of barriers.

[2]As for the motivations, a principal component analysis (PCA) was carried out. It identified three axes—logistical constraints, safety and comfort—explaining 63% of the total variance. The 'bad weather' variable is excluded from the model because it makes only a small contribution to the explanatory axes (which is due to the fact that its influence is linked to both comfort and safety). Nevertheless, given its importance, we will comment on this aspect based on the qualitative material.

levels of ease and cyclists' tactics. Any negative image of cycling within the participant's company or the peer group has a negligible impact.[3]

9.2 Weather Conditions

The biggest hurdle is bad weather: more than half of respondents said they agreed with this. This sensitivity is specific to active mobilities, as their users are in direct contact with the environment. More specifically, rain and high/low temperatures have an impact on cyclists' levels of ease, and winter conditions have safety implications (low light, risk of ice, lack of snow clearance):

What restrains me is heavy rain, snow when it settles and, worst of all, ice. However, neither cold nor heat prevents me from using my bike.

For me, the tram is a complementary means of travel. It replaces cycling in bad weather. I never ride in the city if the road is wet. I wear a suit and I don't want to get it dirty.

In winter months, when it is dark in the morning and in the evening, I take public transport because cycling is too dangerous. Motorists don't see bicycles and I have almost been run over several times despite having equipment adapted for riding at night.

Sensitivity to weather conditions differs between the participants—it should be noted that three quarters claim to cycle all year round—as do the strategies they adopt. While some use other modes of transport in certain weather conditions, others claim that appropriate equipment (clothes, hooded cape, etc.) is enough and that the real barriers lie in a lack of habit, poor-parking conditions (in particular in terms of weather protection, see Sect. 10.2), a lack of snow clearance from cycle routes, and the behaviour of motorists:

No barriers because I only see positive things about cycling, even in the rain. Only large amounts of snow hold me back because motorists have very little control over their vehicles and are more afraid, hence they have a greater risk of hitting us. Otherwise, when it's hot, the bike is airy and, when it's cold, it warms you up!

What stops me in winter when it snows are the roads and paths that have not been cleared of snow. Bicycles no longer have a place on roads and paths in winter. Snow is mostly cleared for cars.

---

[3]Barriers are similar for both conventional cyclists an e-bikers (linear regression; $R^2 = 0.990$). Differences are smaller than 1 point for seven items, including criteria for which the e-bike could have an advantage as it lessens the required effort (carrying goods, issues of sweating and clothing, physical effort due to the topography or distance, exposure to air pollution). A small difference is found in terms of risk of theft (e-bikes are more expensive, but this means that their owner is more likely to find a storage solution before purchase) and image. E-bike users are a little bit more sensitive to weather conditions (54.2% vs. 52.3%), activities before/after work (41.0% vs. 39.1%) and accompanying children (22.6% vs. 20.9%), which may be explained by their longer commute and family situation. The same can be said for safety issues, which shows the biggest difference, with 4.2 points between the two groups: as e-bike users commute further and are more likely to live in suburban and rural areas, they are more likely to have to cohabit with motorised traffic at a high speed.

The only thing is really when it's snowing a lot because it's a death-trap. But that's it. After that, for me, if it rains, I have waterproof trousers, so it's good.

9.3 Logistical Constraints

Logistical constraints relate to the transportation of equipment or packages (47% 'somewhat agree' or 'strongly agree' that this is a barrier) and carrying out other activities (leisure, shopping) before or after work (40%). These can make using a bicycle impractical, especially when the distances involved are large.

Choosing a mode of transport is part of an individual's daily organisation. Mobility brings together the different facets of their lifestyle by connecting the places where their activities take place. Changing mode of transport, therefore, frequently involves adjustments to their daily life and organisational skills:

Using a bicycle requires me to think more about my organisation, in order to focus my need for a motor vehicle on a single day, and do everything on that day.

Thanks to bike to work *I have learned to organise myself better in terms of the things I take with me or leave behind, in terms of the time I take to change. This makes the exercise easier and less tedious. The weight of your bag is still an issue when you have to carry your laptop.*

If you "have" to cycle for bike to work, *you learn to organise yourself so as to overcome certain difficulties that previously prevented you from cycling. Whether this is the weather, transporting equipment, or the supposed time required.*

Accompanying children appears further down in the ranking (21%) but only concerns half of the respondents who live in a family unit. Some parents opt for specific equipment (child seats, trailers, etc.), insofar as they feel that the distance is not too great and the route is safe:

I have to take my child with me to the kindergarten (which is very close to my place of work) in a bicycle trailer. This is the most significant constraint to me cycling to work.

I wouldn't ride a bicycle with a small child in a trailer or in a seat on main roads, or only for very short journeys where we can ride on residential streets (30 km/h zones). It's too dangerous for me, although otherwise I always travel on the main roads when I'm alone.

This last quote highlights the importance of safety, which varies according to the spatial context and type of journey (time, duration, reason, etc.). It appears regularly in the comments.

9.4 Safety

Three elements relate to the safety of the individual and their bicycle. They constitute barriers that are more long term and also more diffuse than the previous ones, which are more ad hoc and associated with specific situations. The significance of infrastructure and coexistence with car traffic is obvious. 36% of participants mention the risk of accidents and the lack of safety. Comments touch upon the intensity of road traffic and motorists' behaviour (lack of respect, aggressiveness, inappropriate driving, use of a smartphone, and even insulting cyclists), which can make cycling a *'constant battle'* and a source of *'fear'*:

I'm generally afraid of motorists who have no regard for others. As a cyclist, I often feel undervalued by motorists. The most dangerous place on my commute to work is in Baden [a German-speaking town]. I'm always frightened there. If I have enough time, I make a big detour so I can do more training and have a safer route in terms of traffic.

The EXCESSIVE speed of car and motorcycle traffic [...] AND the absence of a clearly defined cycle path where no other vehicle [...] can drive.[...] The last roundabout on my journey is located just after a motorway exit. Although I am clearly visible in traffic (fluorescent jacket or vest + light), I get cut off 50% of the time.

I love my bicycle and that's why I use it a lot (2 h 10 min every day) but in Geneva, cycling is a constant battle.

Cycling often requires highly vigilant behaviour:

A car driver looks after himself and just drives. When you're a cyclist or a biker, you have to look after yourself and others, and where other people are [...]. When a car comes up behind us, when we hear the engine, we say to ourselves: what should I do, should I go over to the side, should I move, should I accelerate so as not to annoy them... We have to be constantly listening.

The feeling of being unsafe is also closely linked to a perceived lack of consideration for cyclists and their needs in terms of facilities and infrastructure:

I really don't like it when the yellow bike lanes stop suddenly at traffic lights or at precisely the most tricky points, such as roundabouts, crossings, lane preselections... It is very unsettling. So, for safety, I prefer to cross at pedestrian crossings, pushing my bike along next to me, but that's also annoying for pedestrians.

Nothing restrains me from riding my bike [...]. But in reality I get the jitters every day because infrastructures for bikes are put in place by people who do not ride a bike and they are very dangerous and, most of all, hugely inadequate.

Cohabitation with road traffic encourages the adoption of various tactics, such as careful choice of route (see Sect. 7.2), as well as the use of specific equipment (lights, high visibility clothing, helmets, etc.). The issue of compliance with traffic rules is also raised. In the last comment, the individual prefers to get off their bike in certain places:

I tried travelling to work by bicycle. Unfortunately, the journey is very dangerous in parts because there are no cycle tracks in certain places on the heavily frequented roads and crossroads in Zurich. Not only is car traffic stressful but, unfortunately, so too are other cyclists who carry out very dangerous manoeuvres on the pavement due to the lack of cycle routes.

Other respondents say they break certain rules to avoid potentially dangerous situations. The case of red lights is mentioned; not obeying them enables cyclists to keep a safe distance away from buses, to get ahead of cars which start faster and which sometimes come too close to cyclists or turn just in front of them, to avoid inhaling exhaust gases, etc. Such behaviour is addressed in different ways in the comments. Some regret the image it gives cyclists, others justify it and many call for adaptations to crossroads and their regulation (see Sect. 11.2):

> *Traffic rules and traffic management are designed for cars. No respect for cyclists! It's always the cyclists who have to brake and even make 90-degree turns!!! Never the cars. If cyclists no longer follow the rules, I understand.*

A certain injustice is called out: cycling is promoted due to its beneficial aspects (reducing ecological footprint, fighting against sedentary lifestyles, etc.) but some people consider that the practice itself is marginalised. A tension is thus identified between the societal challenges and the day-to-day experiences on the roads:

> *Cycle routes are awful. While cars have flat asphalt roads, on which they can drive quickly, long-distance cycle routes mostly go up hill and down dale, across fields, ascending, descending, taking detours, and travelling through places that sometimes have paved streets. We make life easy for the people generating the exhaust fumes. Then make it complicated for those who get on their bikes and do something for the environment and their health.*

Next, 26% mention that the risk of theft or vandalism of their bicycle—and ultimately the costs and inconvenience that then fall upon cyclists—constitutes a barrier to cycling. This can be linked to the lack of a secure parking place at home or at the destination (this point is more specifically addressed in Sect. 10.2):

> *The lack of adequate facilities for parking bicycles near public buildings, stations, etc., is a barrier to cycling. 'Adequate' here means, for example, being able to lock your bicycle to a bike rack, near to the entrance, and not in an underground garage.*
>
> *A secure bicycle garage [...] would be useful. [...] My first bike lasted two weeks in the bicycle park. I haven't had the second one stolen yet, but I had to "sacrifice" it, i.e. tag it all over with graffiti when it was new to make it less attractive to thieves.*

Exposure to air pollution is a problem for a quarter of respondents. Cyclists can be directly subjected to the noise and pollution generated by motorised vehicles:

> *Air pollution is in part a very serious issue, if you have to drive behind a line of cars and lorries. If it's on an uphill climb and you have to breathe deeply, it's all the more serious...*

The safety concerns relate to the perception or experience of being vulnerable in comparison to motorised road users, to the lack of dedicated infrastructure and the lack of separation from road traffic (both in terms of speed difference and air pollution). They also refer to a perceived lack of consideration by motorists and a lack of legitimacy in the eyes of planners and politicians. The analyses of the bikeability of journeys (Sect. 10.3), of the evaluation of public policies (Sect. 11.1) and of measures implemented to promote cycling (Sect. 11.2), will confirm the importance of safety issues as a factor limiting the expansion of the practice of cycling.

9.5 Comfort

The other barriers, which are quantitatively less significant, relate to comfort. Between one fifth and one-sixth of respondents mention the problems that can arise from physical exertion, sweating, clothing and the amount of time the journey takes (due to distance or gradient):

The main reason for me not cycling [...], I have to carry other clothes with me for work, including shoes, and when at work I need more time to change and take a shower.

500 m of elevation between my work and my home, 20 min to get there, 45 min to get back...

With the increasing popularity of the electrically assisted bicycle, a larger population is adopting cycling or continuing the practise (see Sect. 6.1). This type of bicycle makes it possible to travel longer distances, to offset some of the elevation, and reduce the effort required:

The too-long journey time and too-great physical effort stop me from using a normal bicycle. With the e- bike, these things are eliminated. These are the reasons that I bought one.

I bought an electrically assisted bike. I don't sweat anymore. I can go to work in normal work clothes and I don't Have to take a shower. With a normal bicycle, I wouldn't do it.

Comments left in the questionnaire also include thoughts around the laziness or lack of will that affect regular cycling, as well as competition from other modes of transport:

The biggest factor holding me back from cycling is my own laziness.

It's so easy to jump in a car! It requires a certain willpower and acceptance to leave the comfort of other modes.

Where I live is very well served by public transport and it's a bit faster.[...] I'm actually discouraged by public transport. The metro is literally twenty metres from my house.

* * *

Knowing the barriers is important with a view to expanding the practice of cycling to other audiences, since newcomers to cycling are likely to perceive these issues as being more difficult to overcome. The barriers relate to certain—obvious but fundamental—characteristics of cycling. Cyclists are in direct interaction with the environment, have no bodywork to protect them, and are propelled by their own muscular energy.

While the sensory experience of the environment and the use of one's own muscular energy both constitute motivations for cycling (see Chap. 8), they are also the basis of the four families of barriers encountered by commuters: weather conditions, logistical constraints, safety concerns and issues of comfort. These barriers are of different natures and have different impacts on the practice of cycling. Many are time contingent, linked to specific circumstances, and do not prevent cycling in general. This is particularly the case with personal obstacles (owning a bicycle that meets the cyclist's needs, having appropriate equipment, having the required level of physical fitness, riding in bad weather, etc.).

Others reflect the territory's hosting potential. More so than climate or topography, safety issues are fundamental. They are more diffuse than the other identified barriers but are more likely to prevent the adoption or continuation of cycling. They highlight both the problems of cohabitation with road traffic and the need for cycling facilities and infrastructure, two points that we address in the next chapter.

References

1. M. Flamm, *Comprendre le choix modal: les déterminants des pratiques modales et des représentations individuelles des moyens de transport* (Ecole Polytechnique Fédérale de Lausanne, Lausanne, 2004)
2. E. Heinen, B. van Wee, K. Maat, Commuting by bicycle: an overview of the literature. Transp. Rev. **30**(1), 59–96 (2010). https://doi.org/10.1080/01441640903187001

Part IV
Territories' Cycling Hosting Potential

Chapter 10
Territories

In what types of spatial context do people most frequently use bicycles for commuting? How do commuters judge the quality of their cycled journeys? What are the most problematic aspects of these routes? Are there differences between regions and cities?

A territory's hosting potential for cycling—or its bikeability—refers to how receptive or well adapted, it is to the practice. In this chapter, we discuss two aspects of hosting potential: spatial context (cantons and types of municipality) and home–work journeys, first from the perspective of parking conditions, then amenities and infrastructure, as well as cohabitation with other road users.

10.1 Marked Spatial Disparities

The distribution of places of residence of *bike to work* participants (Fig. 10.1) eveals first of all that about 250 of them live in France, and around the same number live in Germany. These cross-border workers predominantly cross the border around Geneva and Basel. The location of the centres of employment and the existence of cycling infrastructures allow them to cycle to Switzerland. Other cross-border workers combine cycling and public transport, as is the case for those coming from the German city of Freiburg im Breisgau, to the north of Basel. Finally, around 20 participants are resident in Austria and Liechtenstein.

The map also shows a concentration in German-speaking regions (Eastern Switzerland) and cities that can be observed based on an analysis by canton and type of municipality.

Comparing the proportion of each canton in the sample with their percentage share of the resident population in Switzerland, Table 10.1 highlights an over-

© The Author(s), under exclusive license to Springer Nature Switzerland AG 2021
P. Rérat, *Cycling to Work*,
SpringerBriefs in Applied Sciences and Technology,
https://doi.org/10.1007/978-3-030-62256-5_10

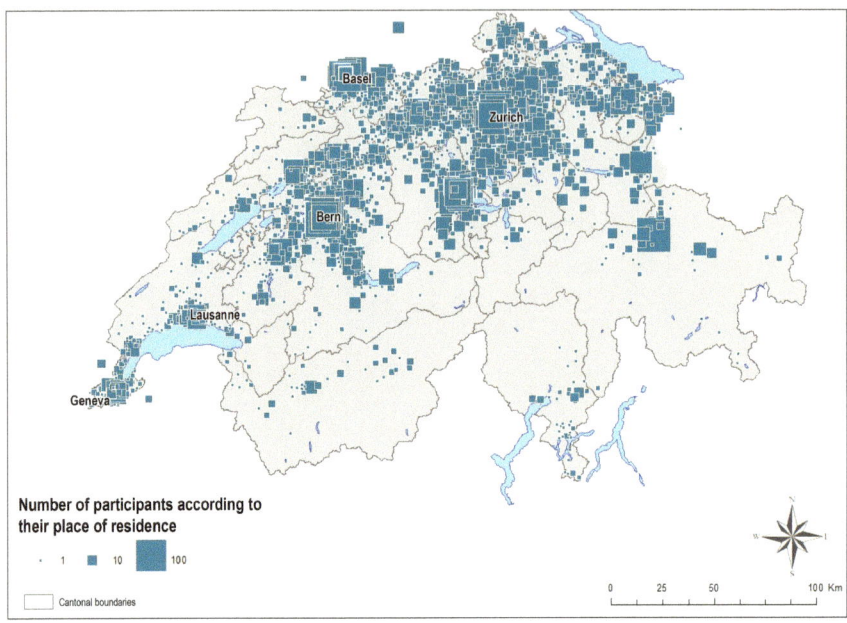

Fig. 10.1 Places of residence of *bike to work* participants according to postcode

representation of the majority of German-speaking cantons.[1] This is true, in particular, for Bern (difference of +9.2 points), Zurich (+6.3), Aargau (+2.1) and Basel-
Stadt (+1.3). Conversely, small, rural German cantons and, above all, Latin cantons,
are under-represented. Vaud, for example, is the fourth most populous canton in the
country, with 9.3% of the Swiss population, but only represents 4.0% of the sample
(−5.3 points). The same trend is observed in Ticino (−3.7), Valais (−3.3), Geneva
(−3.0), Neuchâtel (−1.7) and Fribourg (−1.3). The overall difference between the
German-speaking and Latin cantons reflects different levels of practice (as we saw
in Sect. 3.1). Other factors contribute to this difference: the *bike to work* scheme
is better known in German-speaking Switzerland than in Latin Switzerland, and its
advertisement on social media was almost exclusively in German in 2016. An analysis of the bikeability of journeys and the role of public authorities will shed light
on the causes of this difference.

Breaking the sample down by type of municipality (Table 10.2) shows a clear
over-representation of *bike to work* participants in large urban centres (+10.6)
as well as in large nearby suburban municipalities (+1.9). The other categories
are under-represented, although medium-sized urban centres and their suburbs are
only slightly under-represented. The difference is, however, very pronounced for

---

[1]There is one Italian-speaking canton (Ticino), six cantons with a majority of French-speakers
(Fribourg, Geneva, Jura, Neuchâtel, Valais, Vaud). The other cantons are German-speaking
(including Grisons where Rhaeto-Romanic and Italian are also spoken).

Table 10.1 Proportion of each canton in the sample and in the resident population of Switzerland

	Number of respondents	Proportion of the sample (%)	Proportion of the population (%)	Percentage difference
Zurich	3,003	23.8	17.5	+6.3
Bern	2,699	21.4	12.2	+9.2
Aargau	1,262	10.0	7.9	+2.1
St. Gallen	818	6.5	6.0	+0.5
Lucerne	710	5.6	4.8	+0.8
Vaud	510	4.0	9.3	−5.3
Basel-Landschaft	481	3.8	3.4	+0.4
Basel-Stadt	450	3.6	2.3	+1.3
Solothurn	445	3.5	3.2	+0.3
Geneva	366	2.9	5.9	−3.0
Thurgau	293	2.3	3.2	−0.9
Zug	275	2.2	1.5	+0.7
Fribourg	275	2.2	3.5	−1.3
Grisons	259	2.0	2.3	−0.3
Schwyz	116	0.9	1.9	−1.0
Schaffhausen	110	0.9	1.0	−0.1
Valais	95	0.8	4.1	−3.3
Obwalden	93	0.7	0.4	+0.3
Ticino	79	0.6	4.3	−3.7
Neuchâtel	68	0.5	2.2	−1.7
Glarus	58	0.5	0.5	0.0
Appenzell Ausserrhoden	58	0.5	0.7	−0.2
Nidwalden	48	0.4	0.5	−0.1
Jura	26	0.2	0.9	−0.7
Uri	24	0.2	0.4	−0.2
Appenzell Innerrhoden	15	0.1	0.2	−0.1
SWITZERLAND (not including neighbouring countries)	**12,636**	**100**	**100**	

rural peri-urban municipalities (−4.5), agricultural municipalities (−3.5) and even touristic municipalities (−1.7).

The disparities between types of municipality are explained by different aspects of the territorial structure: density of inhabitants and of jobs, varying mixes of economic functions, distances to be covered, competition from other modes of transport, etc.

Table 10.2 Proportion of each type of municipality in the sample and in the resident population of Switzerland

	Proportion of the sample share (%)	Proportion of the population (%)	Difference
Large urban centres	27.0	16.4	+10.6
Secondary centres of large urban centres	12.6	10.7	+1.9
Suburbs of large urban centres	16.5	17.9	−1.4
Medium urban centres	12.8	13.1	−0.3
Suburbs of medium urban centres	15.6	15.8	−0.2
Small urban centres	1.7	2.6	−0.9
Rural peri-urban municipalities	9.6	14.1	−4.5
Agricultural municipalities	3.6	7.1	−3.5
Tourist towns	0.6	2.3	−1.7
Total	100	100	

The business sectors represented by the companies involved in *bike to work*—mainly higher level services—are less frequently located in rural areas, which may also explain the under-representation of these areas.

10.2   Variable Parking Conditions

Parking represents a crucial factor in a territory's bikeability. Adequate parking conditions make it possible to ensure the continuity and regularity of the practice. If the parking place is not sheltered and secure, it will not protect bicycles against bad weather, or even damage or theft. If it is not easily accessible, it will make using the bicycle more difficult (see Sect. 9.4).

Bicycle storage areas are not specifically provided at more than half of the respondents' place of residence (Table 10.3). However, more than 6 in 10 cyclists use a sheltered and secure place (garage, cellar, apartment or dedicated space). Nevertheless, a significant minority of respondents do not have good conditions for storing their bicycles:

> *My old bike was stolen from outside my apartment. Now, with my new bike, I always take it up to the balcony for fear somebody will steal it. I live on the first floor, so it's not a big deal, but still. I have to take it up and down the stairs. It's a bit of extra effort that puts me off a little.*

Certain differences appear between types of municipality. Residents of large cities are less able to use sheltered places (55.6%) or secure premises (52.0%) than other

Table 10.3 Characteristics of bicycle parking places

	At home (%)	At work (%)	At a public transport stop (%)
Place specifically reserved for bicycle parking	45.8	73.9	84.5
Sheltered location	61.4	66.1	58.0
Place equipped with bicycle racks	12.9	31.6	50.6
Secure premises (e.g. lockable shelter)	62.3	26.4	14.5

respondents. Residents of suburbs and rural communities are less likely to have specific places for parking their bicycles (less than 45%). Given the typical dwellings in these areas (detached or semi-detached houses, etc.), they are often able to store their bicycles in other types of premises (e.g. a garage), which are secure and sheltered from the weather.

The situation is somewhat different at the workplace: three quarters have access to a place for parking bicycles. This can take various forms, such as markings on the ground, a sheltered space such as a canopy (66%), and, more rarely, a secure space such as a shed, garage or other lockable location (26%). Three in 10 participants regularly lock their bicycles to a bicycle rack or similar.

Commuters who leave their bicycles at stations or other public transport stops are more likely to have access to places which are specifically designed for bicycle storage, and to be able to secure them to a suitable structure, such as a bicycle rack. Conversely, they are less likely than the other categories to have access to a lockable bicycle shed or sheltered space.

10.3 Journeys with Inadequate Bikeability

The *bike to work* participants were asked about the bikeability of their commute. The respondents were presented with several statements in respect to their commuting experiences (Fig. 10.2). The results, therefore, do not take into account journeys completed by bicycle for other purposes.

With regards to cycling facilities and infrastructure, three quarters of the respondents state that they do not have continuous cycle lanes on their route:

It comes, it goes, it comes, it goes. And that's even worse than if there was nothing. Because if there were no lines marking out the cycle lane, at least the cars would realise that the bikes are in the same traffic as them.

Cycle lanes often stop where the road narrows. The cyclist is then immediately confronted with individual motorised vehicles without a way of avoiding them.

The lanes intended for bicycles have a broken line. Conveniently, dangerous places (like where roads narrow) do not! I would prefer ten metres of cycle track in a dangerous place, than kilometres of cycle track where there is no danger.

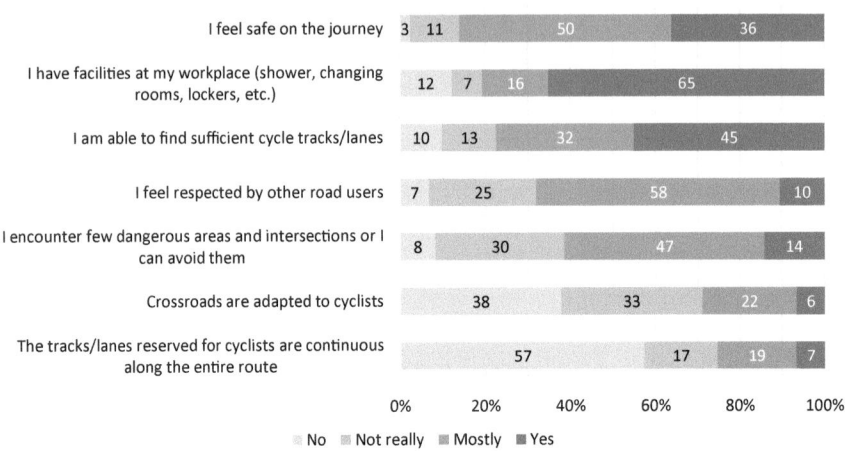

Fig. 10.2 Experience of cycling on the home–work commute

The infrastructures are not continuous! There is no coherence. They've said to themselves, "well, we put a track there and it's all good", as if cyclists can teleport from one track to another.

Three quarters of cyclists say they still find enough cycle lanes or tracks along their commute, which may seem contradictory given the lack of continuous routes. This seems to be mostly explained by the possibility of using routes with limited traffic volumes and reduced speeds. This is the case with 30 km/h zones and residential streets in built-up areas, as well as paths through the countryside:

I go via the shortcuts, the smaller roads that bypass cities, footpaths through fields. The peace, safety, lack of pollution and beauty of the route are guaranteed.

However, potentially dangerous points cannot always be avoided, as mentioned by 38% of respondents. The problem of crossroads re-emerges here: less than 30% of participants consider them to be suitable for cyclists:

As a cyclist, we are not taken seriously at roundabouts. Motorists often try to overtake you, even when you are already in the middle of the road or the roundabout. This often leads to dangerous situations. The same is true with regard to priority on residential streets. Here too, we are not taken seriously and we are often cut up despite having right of way.

These deficiencies often mean that cyclists do not have a positive experience of cohabiting with road traffic. A third of cyclists state that they are not respected by other road users and 14% say that they do not feel safe on their commute:

On a daily basis, there are pedestrians walking along the cycle track. Lorries, cars, motorcycles and scooters are parked on the cycle track. Failure to respect the right of way of cyclists on a cycle track by vehicles joining from a side road. Dangerous overtaking, etc.

I don't feel at all safe cycling outside of my local area. There are many motorists who overtake very close at high speeds, when there is often plenty of space available. Many motorists are unaware of how unsettling and dangerous it is for them to drive so fast next to a bicycle.

These results are closely linked. Indeed, the dangers relating to using bicycles do not arise from the activity itself but rather from the interaction with motorised vehicles [1]. It is not a matter of an inherently dangerous practice, but of a safe activity which is often practised in a dangerous environment.

E-bikers feel less safe than conventional cyclists (16.1% vs. 13.5%) and less respected by other road users (36% vs. 31.1%). Even though they are more likely to drive a car than other cyclists, they are more critical about cohabitation with motorised traffic. Electric assistance does not seem to help cyclists to cope with motorised traffic, although other factors may come into play, such as the length of commuting trips, the spatial context and age and gender.

Feelings of safety and being respected by other road users vary greatly from place to place as well. With regards to safety, at the cantonal level, a ratio of one to eight is observed between the minimum and maximum values (Table 10.4). The percentage of participants stating that they do not feel safe does not exceed 10% in rural or mountainous German-speaking cantons (Nidwalden, Uri, Obwalden, Graubünden and Glarus). Certain urban cantons, Basel-Stadt and Zug, are also below this threshold. In contrast, cyclists who do not feel safe are over-represented in the Latin cantons: Vaud (34%), Ticino (33%), Fribourg (30%), Geneva (24%) and Neuchâtel (21%).

Table 10.4 Feeling of safety on the home–work commute by canton of residence

Ranking	Canton	% of 'no' and 'not really' responses (%)	Ranking	Canton	% of 'no' and 'not really' responses (%)
1	Nidwalden*	4.2	14	Lucerne	12.5
2	Uri*	4.2	15	Bern	12.6
3	Obwalden	5.5	16	Zurich	14.3
4	Grisons	5.6	17	Schwyz	14.8
5	Zug	7.4	18	Jura*	16.0
6	Basel-Stadt	7.9	19	Valais	17.0
7	Glarus	8.6	20	Appenzell Ausserrhoden	17.2
8	Schaffhausen	9.4	21	Neuchâtel	20.6
9	Basel-Landschaft	10.1	22	Appenzell Innerrhoden*	23.1
10	Thurgau	10.3	23	Geneva	24.1
11	Solothurn	10.4	24	Fribourg	29.5
12	Aargau	10.9	25	Ticino	33.3
13	St. Gallen	11.4	26	Vaud	34.1

Note The number of responses is less than 50 for four cantons: Nidwalden, Jura, Uri and Appenzell Innerrhoden

Table 10.5 Feeling of safety on the home–work commute by city of residence

Ranking	City	% of 'no' and 'not really' responses (%)	Ranking	City	% of 'no' and 'not really' responses (%)
1	Zug	4.4	13	Bern	13.2
2	Burgdorf	4.9	14	Lucerne	13.8
3	Chur	5.2	15	Uster	14.0
4	Baar	6.1	16	Wettingen	14.5
5	Solothurn	6.3	17	Köniz	14.5
6	Winterthur	6.4	18	Baden	15.9
7	Rapperswil-Jona	7.8	19	Biel	18.9
8	Basel	7.9	20	Fribourg	21.8
9	Aarau	8.9	21	Ostermundigen	21.8
10	Thun	9.4	22	Zurich	22.3
11	Kriens	10.8	23	Geneva	22.4
12	St. Gallen	11.6	24	Lausanne	34.2

The differences are also considerable between cities,[2] and the same ratio of one to eight is observed (Table 10.5). In the cities of Zug, Burgdorf and Chur, 1 in 20 cyclists does not feel safe during their commute. This proportion is more than four times higher in Fribourg, Ostermundigen, Zurich and Geneva, where more than a fifth of bicycle commuters feel (mostly) unsafe. Lausanne brings up the rear with the maximum value of 34%.

The feeling of cyclists not being respected by other road users also differs between cantons and cities but in lesser proportions (a ratio of one to four). In several cantons, between a quarter and a fifth of bicycle commuters do not feel respected (Table 10.6). These are predominantly rural cantons (Uri, Grisons, Obwalden, Appenzell Innerrhoden) and small urban cantons (Zug, Basel-Stadt). Three French-speaking cantons fall in the middle of the ranking (Neuchâtel, Valais and Jura) and fluctuate around 25%. The others—Fribourg (45.0%), Geneva (51.5%) and Vaud (51.3%)—bring up the rear along with Zurich (34.6%) and Ticino (49.4%).

In terms of the cities (Table 10.7), Burgdorf is notable for having the lowest value (13.3%), followed by Baar, Chur, Basel and Zug at seven percentage points higher. More than 4 in 10 cyclists in Wettingen, Biel, Zurich and Kriens express a

---

[2]The municipalities included in this table are those for which a minimum of 50 responses were obtained for the three questions analysed in this regard: feeling of safety, feeling of being respected by other road users, and satisfaction with the commitment of public authorities to cycling in their region. Some Latin centres—despite their size—are missing from the ranking. Among the 23 municipalities that have more than 30,000 inhabitants, only one German-speaking municipality is missing (Emmen in the suburbs of Lucerne), while the two urban centres in Ticino (Lugano and Bellinzona) and the French-speaking cities of La Chaux-de-Fonds, Vernier, Sion, Neuchâtel and Lancy are not included. German-speaking cities are more highly represented, even when they have relatively modest demographic weight, such as Solothurn or Burgdorf (16,000 inhabitants).

Table 10.6 Feeling of being respected by other road users on the home–work commute according to canton of residence

Ranking	Canton	% of 'no' and 'not really' responses (%)	Ranking	Canton	% of 'no' and 'not really' responses (%)
1	Uri*	8.3	14	Schwyz	28.9
2	Grisons	19.0	15	Aargau	29.2
3	Obwalden	19.8	16	Nidwalden*	29.2
4	Zug	22.7	17	Bern	29.4
5	Basel-Stadt	22.7	18	Basel-Landschaft	30.2
6	Appenzell Innerrhoden*	23.1	19	Thurgau	30.4
7	Glarus	23.2	20	Lucerne	30.6
8	Schaffhausen	24.8	21	Appenzell Ausserrhoden	32.8
9	Neuchâtel	25.0	22	Zurich	34.6
10	Valais	25.8	23	Fribourg	45.0
11	St. Gallen	26.3	24	Ticino	49.4
12	Jura*	28.0	25	Vaud	51.3
13	Solothurn	28.8	26	Geneva	51.5

Note The number of responses is less than 50 for four cantons: Nidwalden, Jura, Uri and Appenzell Innerrhoden

Table 10.7 Feeling of being respected by other road users on the home–work commute according to city of residence

Ranking	City	% of 'no' and 'not really' responses (%)	Ranking	City	% of 'no' and 'not really' responses (%)
1	Burgdorf	13.3	13	Bern	29.8
2	Baar	20.0	14	Uster	32.9
3	Chur	20.2	15	Baden	34.9
4	Basel	20.6	16	St. Gallen	35.0
5	Zug	20.9	17	Ostermundigen	38.9
6	Winterthur	22.1	18	Wettingen	40.3
7	Aarau	26.0	19	Biel	40.6
8	Thun	27.4	20	Zurich	41.0
9	Köniz	28.6	21	Kriens	43.4
10	Solothurn	29.0	22	Fribourg	45.5
11	Rapperswil-Jona	29.2	23	Geneva	50.0
12	Lucerne	29.6	24	Lausanne	55.3

negative opinion regarding this issue. Three French-speaking cities bring up the rear: Fribourg, Geneva and Lausanne. In the latter two of these, 50% and 55% of cyclists, respectively, do not feel respected by other road users.

Feeling unsafe and having to cohabit with road traffic encourage the wearing of helmets among participants. This practice is widespread among the majority of the surveyed population, as we saw in Sect. 6.1.

* * *

The spatial distribution of *bike to work* participants shows an over-representation of residents of German-speaking Switzerland and urban centres, while the inhabitants of small rural German cantons and, especially, Latin cantons, are under-represented. Although certain specificities of the *bike to work* scheme should be noted (location of the types of companies that participate in the scheme, etc.), these results reflect the different levels of utility cycling in Switzerland.

The analysis of home–work journeys shows gaps in the bikeability of the territories, whether in terms of secure and sheltered parking for bikes, the presence of cycle routes or the development of crossroads that accommodate cyclists. Due to the lack of cycling infrastructures and facilities, cyclists frequently have to share space with road traffic in unsuitable locations, and this can be a negative experience. One in three cyclists does not feel respected by other road users, and one in seven states that they do not feel safe on their commute. These proportions are all the more significant given that the population under study is mainly made up of regular cyclists. In all likelihood, these problems are even more acute for people who are less motivated or less competent at cycling.

Marked differences are observed within the country in terms of the *bike to work* participants' feelings of travelling safely or being respected by other road users. The difference in practice between German-speaking Switzerland and Latin Switzerland is often interpreted as the consequence of cultural differences. It appears here to be primarily a consequence of very different traffic conditions. Participants' dissatisfaction with transportation and planning policies is discussed in the next chapter.

Reference

J.R. Pucher, R. Buehler (Eds.), *City Cycling* (MIT Press, Cambridge, 2012)

Chapter 11
Politics

How do cyclists assess public authority commitment to cycling in their region? What territorial disparities can we see? What measures are suggested by the interviewees?

This chapter builds on the previous one by approaching the bikeability of territories from a political angle. It focuses on the actions of public authorities as perceived by the bicycle commuters and then discusses the measures recommended by participants to promote cycling.

11.1 Critical Appraisal of Public Authority Action

The actions taken by the public authorities to promote cycling are an indicator of the legitimacy—both political and social—attributed to this mode of transport. The surveyed population's evaluation of the public authorities' strategies is highly mixed. Almost 50% believe that the authorities in their region are not sufficiently committed to cycling (17% no and 33% not really) and the other half believes the opposite (8% yes and 41% mostly).

The level of dissatisfaction varies greatly depending on the territory. In general, the German-speaking Swiss population is less dissatisfied (48.4% no or not really) than French (65.3%) and Italian speakers (72.6%). Ticino has the second most negative responses for this value after Fribourg (74.5%) (Table 11.1). They are followed by a number of cantons where two-thirds express dissatisfaction: Vaud, Schwyz, Neuchâtel, the two cantons of Appenzell, Obwalden and Valais. In only five cantons do more than 60% of the respondents believe that the public authorities take sufficient action with respect to cycling: Glarus and Schaffhausen (but both in limited numbers), Zug and the two cantons of Basel, where the lowest numbers of negative responses are observed (34.6 and 29.1%) (Table 11.2).

© The Author(s), under exclusive license to Springer Nature Switzerland AG 2021
P. Rérat, *Cycling to Work*,
SpringerBriefs in Applied Sciences and Technology,
https://doi.org/10.1007/978-3-030-62256-5_11

Table 11.1 Perception of sufficient commitment to cycling by public authorities according to canton

Ranking	Canton	% of 'no' and 'not really' responses (%)	Ranking	Canton	% of 'no' and 'not really' responses (%)
1	Basel-Stadt	29.1	14	Geneva	53.7
2	Basel-Landschaft	34.6	15	Lucerne	54.4
3	Schaffhausen	35.0	16	Nidwalden*	57.8
4	Glarus	35.8	17	Jura*	59.1
5	Zug	38.6	18	Valais	60.2
6	Grisons	43.8	19	Obwalden	61.4
7	Thurgau	46.9	20	Appenzell Innerrhoden*	61.5
8	Bern	47.0	21	Appenzell Ausserrhoden	65.5
9	St. Gallen	47.5	22	Neuchâtel	69.0
10	Solothurn	48.5	23	Schwyz	69.5
11	Aargau	48.5	24	Vaud	70.5
12	Uri*	52.4	25	Ticino	72.6
13	Zurich	53.0	26	Fribourg	74.5

Note The number of responses is less than 50 for four cantons: Nidwalden, Jura, Uri and Appenzell Innerrhoden

Table 11.2 Perception of sufficient commitment to cycling by public authorities according to city

Ranking	City	% of 'no' and 'not really' responses (%)	Ranking	City	% of 'no' and 'not really' responses (%)
1	Winterthur	23.5	13	Zug	50.8
2	Burgdorf	29.3	14	Thun	52.2
3	Basel	29.6	15	Uster	53.3
4	Baar	34.0	16	Baden	53.3
5	Chur	35.2	17	Kriens	53.4
6	Solothurn	38.6	18	Lucerne	54.8
7	Ostermundigen	43.1	19	Wettingen	56.1
8	Aarau	44.9	20	Geneva	56.8
9	Köniz	45.5	21	St. Gallen	58.4
10	Bern	46.1	22	Zurich	65.6
11	Biel	47.3	23	Lausanne	70.3
12	Rapperswil-Jona	47.5	24	Fribourg	76.9

Large disparities are once again found at the city level. Results are taken into account for cities for which a minimum of 50 responses were obtained. Three French-speaking cities appear among the five cities with the highest proportions of negative responses. The majority of cyclists who live there believe that the public authorities do not show sufficient commitment to cycling: Fribourg (76.9%, the highest value recorded), Lausanne (70.3%) and Geneva (56.8%). Size does not seem to be a factor in the order of appearance in the ranking. Some large cities are well ranked—Winterthur (1st place; less than one cyclist in four is dissatisfied), Basel (3rd) and even Bern (10th)—while others are towards the bottom of the ranking—Lucerne (18th), St. Gallen (21st) and Zurich (22nd). This last case is notable: in Zurich, strong growth is observed in the number of cyclists (see Sect. 3.2), but this seems to be independent of the infrastructure available. We also find small and medium-sized cities at the top of the ranking (Burgdorf, Chur, Solothurn, Aarau), as well as suburban municipalities (Baar, Ostermundigen, Köniz).

A high correlation is observed between the feeling of safety on the commute and the assessment of how committed the public authorities are to cycling (Fig. 11.1): the less safe the commuters feel in their movements, the less they are satisfied with the actions of the public authorities. The latter are blamed for not showing sufficient consideration of the needs of cyclists, whether in terms of infrastructure or regulation of motorised traffic. One might expect that cyclists who live in regions where cycling is heavily promoted would be more demanding, as they are accustomed to specific infrastructure. This does not seem to be the case.

Among the cantons where cyclists feel the safest, we can identify small urban cantons on the one hand (where cyclists are also more satisfied with the public

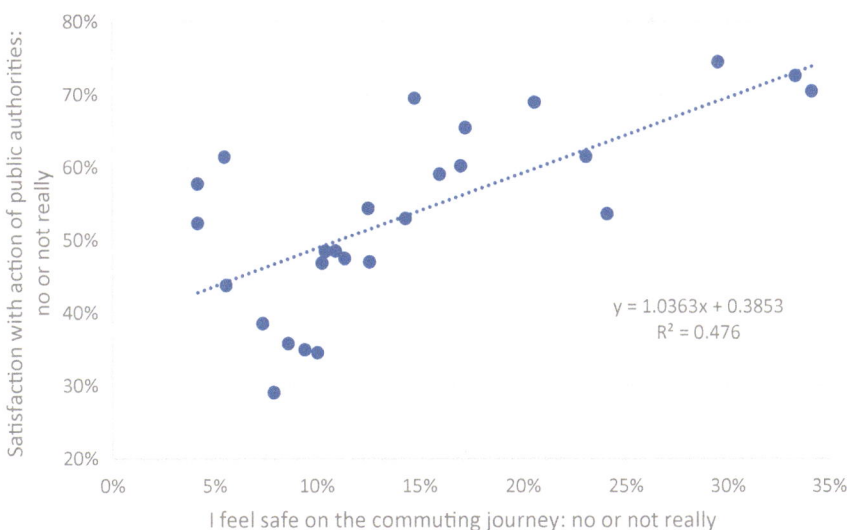

Fig. 11.1 Relationship between feelings of safety during the commute and perceptions of the commitment of public authorities (canton level)

authorities)—Basel-Stadt, Basel-Landschaft, Zug and even Schaffhausen—and, on the other hand, small rural cantons—Uri, Nidwalden, Obwalden—which benefit from being located away from the main traffic routes, but where commuters are more critical.

As we have already mentioned, the variable level of cycling between cantons is sometimes interpreted from the perspective of cultural differences between German-speaking Switzerland, which is more inclined to cycle, and the Latin regions, which are more reluctant due to an attachment to the car. These results show, however, that a substantial proportion of the differences results from political issues. When the authorities give little weight to cycling, its position within the transport system is reduced, feelings of safety are undermined, and the modal share remains limited.

11.2   The Measures Recommended Most

Feelings of being unsafe, lack of adequate infrastructure and dissatisfaction with the public authorities explain the very large number of comments (4,404) relating to measures recommended by respondents to promote cycling on their commute. This figure is all the more significant given that the comments were in response to an open question at the end of a long questionnaire. The comments were processed in order to divide them by type of measure suggested, and on the basis of the theoretical framework presented in Chapter 2, nine main types were identified. As a single comment can address multiple types of measure, a total of 6,993 occurrences were noted (Table 11.3). Promotion and awareness-raising measures address action levers linked to individual mobility potential (access, skills, appropriation). The most frequent suggestions relate to the territory's hosting potential, whether in terms of infrastructure and amenities, or social and legal norms and rules. We will comment on them in the order of frequency with which they appear.

11.2.1   Safe and Developed Infrastructure

Improvements to infrastructure are suggested most commonly and represent two-thirds of the measures discussed. The types of measure suggested most frequently (44%) concern routes, and more particularly their separation from road traffic. This clearly reflects the need for safety. The development of cycle lanes and tracks is mentioned frequently. As we observed in the section on levels of ease (see Chapter 7.1), cycle tracks are considered more effective than cycle lanes, which are sometimes seen as unworkable and insufficiently safe:

All roads on which cars can travel at more than 50 km/h should have a separate cycle track, not just a cycle lane.

Real cycle tracks and not a few paint strokes on the road to show that there are however many kilometres of cycle routes in the canton! Routes off main roads.

Table 11.3 Typology of measures recommended by respondents to promote cycling on the home–work commute

Theme	Suggested measure	Number of occurrences	As a percentage of the total (%)
Promotion and awareness raising		442	6.3
Infrastructure and amenities		4,540	64.9
	Routes	3,043	43.5
	Crossroads	540	7.7
	Parking	437	6.2
	Improved route layout	433	6.2
	Connection with public transport	87	1.2
Norms and rules		1,680	24
	Traffic rules	853	12.2
	Improvements with regards to coexistence with road traffic	695	9.9
	Better recognition of cycling	132	1.9
Other		220	3.1
No measures required		111	1.6
Total number of suggestions		**6,993**	**100**

We need separate bike routes. When I cycle to work, I feel welcome neither on the 80 km/h road, nor on the bike route, nor on the paths which I have to share with pedestrians, runners, dogs, etc.!

What would reassure me is not so much boundary markings on the ground, but rather a real physical boundary. Because I get the impression that even me, when I am driving a car, it is sometimes difficult to share the road with bikes because there aren't necessarily places for them. Sometimes their actions are a little unpredictable […] Whereas with a physical boundary, it would be better.

Another criterion is the continuity of cycle routes, which are frequently interrupted, according to the participants. Measures should be taken to make cycling not only safer but also more efficient (build express cycle routes, fill gaps in the network, etc.):

Fluid, safe and uninterrupted cycle tracks all the way along the road, with right of way: for efficient transport, without too many detours.

The authorities often simply offer alternative routes, which are often longer and less direct, rather than developing more direct stretches of safe cycle lanes.

A continuous network of cycle routes should be built. The proposal must come from the canton. Otherwise there are too many disputes between municipalities. Some municipalities are very positive and others don't think it's necessary to build cycle routes at all.

Other comments, albeit with much less frequency, relate to intermodality, and specifically to coordination with public transport networks:

Being able to load a few bikes onto buses more easily. The trains are also very poorly suited to bikes. They should also remove the requirement to pack up your folding bike on trains!

Improving transportation of bikes on the train. It has got worse since 2008 with the removal of bicycle hooks from train carriages. More bicycle parking at stations, especially for long bikes such as cargo bikes, which make it much easier to transport children, if they have to be taken to nursery before work.

Three other elements relating to infrastructure make up 6–8% of the suggested measures. First, crossroads appear to be potentially dangerous places. This relates in particular to roundabouts, where it should be remembered that one-third of respondents do not feel (very) at ease (see Chapter 7.1). The comments are often general but also relate to specific places, and some are even georeferenced, which is a testimony to the problematic situations experienced on a daily basis. According to the comments received, crossroads should be better designed so as to show more consideration for cyclists (not interrupting cycle lanes at crossroads, planning alternative routes, advanced stop lines, etc.):

At large crossroads and roundabouts, cycle lanes could be better integrated, or even just exist full stop.

Better signage for cyclists at roundabouts. It's time to run an awareness campaign on the BICYCLE AT ROUNDABOUTS. Many motorists do not know that bicycles are equal to cars at roundabouts. This results in very dangerous situations!

Fewer roundabouts!

Second, parking can be problematic if it does not allow bicycles to be stored near to the desired location in a secure (protected against theft and damage) and sheltered (from bad weather) manner. The suggested measures are both quantitative (providing enough spaces, solving the problem of abandoned bikes, etc.) and qualitative (configuration, equipment, etc.) and also relate to location (proximity, accessibility, etc.):

Bicycle parking spaces, especially for homes. In my case, I have to haul my bike up from my cellar because there is nowhere to park it securely where I live.

I would like to park my bike in a bike shed instead of having to leave it on the street. It has already been damaged once out on the street.

More bicycle parking spaces where you can attach both the rear wheel and the frame to a fixed point. Places where you can only lock the rear wheel to the frame of the bike are not safe enough.

Finally, route layout is mentioned. This refers to signage, surfacing, lighting and even maintenance, the quality of which could be improved. This category also includes comments regarding construction sites (lack of detours or alternative routes for cyclists, traffic signs obstructing the cycle lane, etc.):

Depending on the place, the edges of roads are often in very bad condition. Potholes, puddles that freeze in winter, poor transitions between surfaces due to road works, etc.

Better snow clearance in winter. Often cycle paths are used to store snow which has been cleared from the road.

Better handling of cycle lanes and signage. The cycle network is out of date!

11.2.2   Norms and Rules Legitimising Cycling

After infrastructure, a second set of comments relates to norms and rules. This is, as it were, the *software*, i.e. the way of managing and using the infrastructure that makes up the *hardware*. More than 1 in 10 measures mentioned relate to traffic rules, including direct measures (such as the turn-right-on-red policy, which was frequently mentioned, green waves for cyclists, detection of bicycles by traffic lights, and contra-flow cycling) and indirect measures (such as speed limits for motorised traffic):

Adaptation of the Road Traffic Act concerning compliance with red lights. At certain intersections, and depending on the situation, cyclists should be allowed to overtake or turn at a red light.

Earlier traffic signals for bicycles at crossroads. Starting at the same time as cars and scooters often involves significant danger and a large amount of stress because drivers (car, motorcycle and scooter) do not respect bikes.

Green wave for cyclists like in Denmark. Anyone entering the city at a speed of 20km/h always has green lights and can therefore get across without stopping.

At most traffic lights, cyclists are not picked up as road users. That's why I often have to wait several minutes until a car comes along. Buttons, like the ones in Amsterdam, for example, would reduce the waiting time at crossroads because the embedded sensors at "smart" crossroads don't work.

Problems related to coexistence—with motorised traffic, but also with pedestrians and other bicycle users, especially electrically assisted bicycles—appear in a similar proportion of comments. They relate to respect for rights of way, overtaking distances, sites reserved for bicycles, speed limits, etc. These comments frequently refer to safety issues and are combined with suggestions for improvements that would reduce conflicts with other road users:

I want there to be more fines issued to vehicles parked on cycle paths and for pedestrians to also show a little respect for cycle routes.

Awareness-raising for motorists about sharing the road, courtesy and respect for cyclists. Not encroaching on the cycle lane when there is no need. Not cutting cyclists up and respecting their right of way. Leaving enough distance when overtaking a cyclist.

Many militant cyclists damage the image of cyclists and destroy the work of the cycling lobby. I think it would be good if we didn't just address motorists with the demand for more respect, but if we also held cyclists who do not respect the rules to account.

Some respondents complained about the lack of recognition and consideration of cycling as a means of transport—and not just a leisure activity—at the political level but also, more generally, from the perspective of social and cultural norms. Mention is often made of the countries of Northern Europe, which are viewed as a source of inspiration:

> *Developments for bicycles are carried out by people who do not ride a bike and are therefore highly dangerous and, above all, largely insufficient.*
>
> *The municipal officials in charge of mobility [...] are not at all aware that cycling is a means of transport and not just a leisure activity. Their approach is totally disconnected from the needs of cycle mobility, which is to be able to get from point A to point B quickly and safely.*
>
> *A transport system which considers cyclists equivalent to motorised traffic! Model: The Netherlands! :-)*
>
> *Policies and the public should be much more positive about cycling, not just during bike to work! Routes in Switzerland do not in any way meet the standard of other countries. Policies only reiterate that more use should be made of cycling, but doesn't give much priority or money to infrastructure.*
>
> *Swiss policy does not treat the bicycle as part of mobility in its own right or as part of the solution to address road traffic issues.*

Slightly more than 6% of the suggested measures related to promotional actions (subsidies for purchasing electrically assisted bicycles, tax deductions, development of bicycle sharing schemes, awareness campaigns particularly targeted at young adults, cycling lessons, etc.). Other comments—although rarer—concerned facilities at the workplace (showers, lockers, etc.), the use of interactive maps, etc. Finally, in 111 comments (1.6%), participants did not feel the need to suggest any measures, as they are satisfied with their journey from home to work.

* * *

When asked about cycling policy in their region, half of the *bike to work* participants are critical. The differences between regions and cantons are, however, very large. The highest shares of people who are dissatisfied with the public authorities are observed in the cantons and urban centres of Latin Switzerland.

The measures recommended by cyclists regarding the bikeability of their commute address two main themes: improving infrastructure (*hardware*) and the rules and norms relating to cycling (*software*). These measures should make it possible to improve the experience of riding a bicycle, particularly from a safety perspective (thanks to dedicated routes, separated from road traffic). Underlying these measures is the demand to legitimise the place of cycling within the mobility system and to consider it as a means of transport in its own right.

Conclusion

This research analysed the different facets of utility cycling in Switzerland, using the example of commuting. We took as our starting point the concept of the cycling system, or velomobility, which underlines the importance of taking into account all elements—not only material and technical but also social, political and symbolic—which influence this practice. From this perspective, we argued that cycling—in terms of volume, frequency, distance, motivation, etc.—depends on the coming together of two potentials. The first of these is motility [11–13] or, more precisely, the individuals' cycling potential. It is built around access ('to be able to' use a means of transport), skills (('to know how to' cycle for utility reasons) and appropriation ('to want to' cycle). Individuals' appropriation of cycling depends on their perception of that mode and of its particularities, which can be interpreted as a confluence of three fundamental dimensions of mobility: movement, meaning and experience in a context of power in regards to the dominant system of automobility [6]. The second of the two potentials is the territory's hosting potential, or its degree of bikeability, which relates to the spatial context, the available infrastructure and amenities (bicycle urbanism), as well as social and legal norms and rules.

In order to identify a large sample of bicycle commuters, we focused on the *bike to work* scheme, which each year brings together people who commit to cycling to their place of work as often as possible during the months of May and/or June. Nearly 14,000 people completed an online questionnaire addressing the dimensions of velomobility. This database does not take into consideration those people who are not economically active (the sample includes no children, homemakers or retirees and very few students) or who cycle for other reasons (in particular, leisure or sport). The data nonetheless allow us to better understand utility cycling and to unpack and analyse the various dimensions.

Switzerland is very much mid-table in comparison with other Western countries; the modal share of cycling (around 7%) is higher than that observed in Latin and English-speaking countries, but lower than that observed in the North of Europe [17]. In recent years, the practice of cycling has become more urban. Substantial growth rates can be observed within large and medium-sized cities, although many disparities still exist between them. Cycling has also become a political objective, and cities and cantons alike are beginning to develop action plans to promote the practice. In a public vote held in September 2018, about three quarters of voters elected to include the—non-binding—principle of promoting cycling in the federal constitution. The present study makes it possible to extend these debates by proposing an in-depth analysis of utility cycling. The conclusion brings together a summary of the empirical chapters, and identifies the key lessons that can be taken away from this research.

The Wide Variety of Uses of Utility Cycling

The participants in *bike to work* are characterised by an over-representation of the middle age categories, graduates, city dwellers, and residents of German-speaking Switzerland. This is explained by the varying propensities to participate in *bike to work* (depending on the size and type of business, the location of the places of employment and residence, the importance accorded to cycling in the different territories, etc.) and also to practice utility cycling. A great diversity is observed in terms of use as well as in terms of equipment, motivations and barriers.

Cycle trips are often understood to involve travel, which takes place over a limited distance, and which is only undertaken in good weather. In reality, bicycle usage is much more diverse. Six in 10 participants cycle to work the majority of the time, while one-tenth see it as an opportunity to try out utility cycling. Less than a quarter of respondents only cycle in summer, and the data show that the seasonality of the practice is correlated with the length of the commute.

The distances covered are relatively long, which can be explained by the challenge presented by the scheme but also by the emergence of 'long-distance' bike commuters and the growth in the number of e-bikes. The traditional threshold of 3–5 kms that is often mentioned in planning in Switzerland is clearly to be redefined.

For the majority of participants, cycling is an integral part of their mobility. A typology of cyclists in six categories—systematic, leisure, individual, utility, commuter and occasional cyclists—shows various combinations in terms of frequencies and reasons for cycling. The likelihood of belonging to these categories varies between social groups, as illustrated by differences in gender, age, type of household and residential location.

Another trend that increases cycling potential in terms of territorial coverage is intermodality, i.e. the combination of cycling with public transport, which is practised in different ways by around 15% of the sample.

More generally, the population under study is diverse in its use of different means of transport. One-tenth of participants can be described as exclusive cyclists; they do not have access to a car or a public transport pass. However, the majority are multimodal in the sense that they combine different means of travel depending on the activities they are engaged in or on the time of year. Public transport is the most common other form of mobility; the number of annual public transport passes is higher among the participants than among the Swiss working population. Conversely, the motorisation rate is much lower, despite almost all of the cyclists having a driving licence.

The above results demonstrate the need to put the term 'cyclist' into context, or at least to use it with caution [19]. Indeed, this term ignores the diversity of people who use bicycles for travel and tends to restrict the identity of a group of people to a means of transport. The variety of uses but also of motivations and barriers, as well as the emergence or strengthening of certain trends (increasing popularity of electrically assisted bicycles, combination with public transport, etc.), are fundamental variables. They offer new potentialities to be integrated into the policies promoting active mobility and territorial development. These results finally echo the calls to take into account the diversity of cyclists in terms of both research and policy (planning and promotion). It is necessary to disaggregate the analysis and to identify groups with similar practices in order to reveal the various ways in which people use bicycles and to better understand the factors and needs associated with each group [8, 16].

Cycling to Work: Not Only a Utilitarian Movement but also an Embodiment of Meanings and Experiences that Constitute Crucial

Bicycle commuters highlight three main ranges of motivation. The first refers to well-being: the benefits of cycling can be physical (doing exercise, keeping fit) and mental (disconnection from work, pleasure, sensory experience, etc.). For some, commuting by bike is a way to squeeze an enjoyable and physical activity into a daily routine characterised by time constraints. While less present in the literature on bicycle commuting, the experience of cycling—mediated through the senses— stands out as being a crucial motivation in which the commuting time is seen not as wasted but as valuable. The pleasure linked to the experience of riding a bike refers to a 'hedonistic sustainability', in the words of Bjarke Ingels [9].

The second type of motivation is independence. This relates to certain practical elements of cycling (freedom and flexibility) compared with the constraints of other modes of transportation (congestion, timetables, etc.). The third body of motivations is civic engagement. Cycling here embodies citizenship and is a way to promote respect for the environment on a global scale (in a context of climate change) as well as the local scale (reclaiming the quality of city life and reconnecting with the environment).

The first two of these ranges of commuters' motivations are intrinsic but intersect political issues. For example, exercise relates to public health issues, and cycling may alleviate traffic and public transport congestion and reduce the emission of pollutants and greenhouse gases. Cyclists are, here again, far from being a homogenous group and give a varying degree of importance to the three ranges of motivations mentioned above. A part of this diversity is highlighted by the four categories I identified among Swiss commuters: active, civic, individualist and enthusiast cyclists. These groups are receptive to varying combinations of arguments, which are in turn explained by gender, life course position and residential location (other variables are likely to have an effect but could not be tested, such as attitudes, physical condition or lifestyle).

Both the practice of the conventional bike and of the e-bike refer to similar motivations. It might have been expected that criteria such as respect for the environment, doing exercise or saving money would be rated lower for e-bikers, and time-saving higher. But this is not the case. This could be explained by the fact that e-bikers are more likely to have a car and thus to compare cycling with travelling by car. Thus by reaching groups (couples with children, people in the second part of their career) and spaces (suburban and rural areas) that are more motorised, the e-bike expands the practice of cycling as a complement or alternative to automobility.

This study is based on a population of users mainly employed in the service sector and in a country with an intermediate level of modal share of cycling in comparison with other Western countries. To what extent can these results be generalised? Some trends may change according to participants' profiles (e.g. students or low-income workers are likely to place more importance on money-saving issues) and the context (efficiency might be mentioned more by those in regions with a mature culture of cycling, while civic engagement may be more present where cycling is not yet recognised as a fully fledged mode of transportation). More research—both quantitative and qualitative—would be needed in order to go beyond this case study and address these issues for a variety of population groups, spatial contexts and cycling practices (e.g. bike share). Children and teenagers would be an interesting group to study, as they are the only age group for which cycling is declining noticeably in Switzerland [18]. Other groups could be taken into account, such as students, homemakers and the elderly, as well as people who do not yet use the bicycle as a means of transport but who may be contemplating the idea. Such analysis could not only be cross-sectional but also longitudinal, allowing the observation of potential changes over participants' life courses and cycling trajectories [10, 15].

The e-bike: Extending the Practice of Cycling

The cyclists are predominantly equipped with traditional bicycles (9 out of 10 participants have at least one), although the increasing popularity of electrically assisted bicycles is clearly observable (almost a fifth of respondents own one). The comparison between conventional cyclists and e-bikers enables us to identify to what extent and in which ways the e-bike expands or redefines the practice of cycling [2, 4, 7]. Such questions are important, as sales of e-bikes are booming and might soon reach the number of mechanical bikes sold in several European countries. The e-bike is therefore likely to become an integral part of future sustainable mobility systems.

Electric assistance makes it possible to broaden the practice of cycling to more people. In the present study, women are overrepresented among e-bike users in comparison to conventional cyclists, but the fact that several studies have found the opposite to be the case [14, 20] may indicate a diversification of e-bike adopters over time. People with a lower physical condition, as well as employees in the second stage of their working career, are more likely to adopt the e-bike. However, this study shows that the e-bike is far from being restricted to these groups and is used by younger and fitter cyclists. E-bikers cover on average longer commuting distances and enhance the carrying capacity (cargo function) of the bike, as highlighted by the overrepresentation of people with children.

E-bikes also expand the practice of cycling across different types of space. While conventional cycling is prevalent in urban areas, it may be less attractive to suburban and rural dwellers, perhaps due to the fact that their average commuting journey is longer. Thus the e-bike not only helps to overcome the physical limitations of the user but also makes cycling a possibility for those living further from the workplace, enabling them to cover longer distances than with a mechanical bike. It also appears that the e-bike helps to 'flatten' the topography of a route for its user (although the results of the survey are not geographically precise enough to confirm this).

On the whole, the survey shows that the electric assistance of the e-bike, by diminishing the physical effort required to cycle, improving the carrying capacity and increasing the potential distance travelled, empowers more people to cycle. It expands the practice of cycling across social groups (gender, age and life course position, physical condition) and spaces (e.g. different residential contexts and distances), with the potential to widen the cycling renaissance beyond central areas to other types of space. Both cyclists who use conventional bikes and those who ride e-bikes face, however, common challenges in a country where utility cycling is still a minority practice.

The Restriction of Inadequate Bikeability on Cycling Practices

The barriers pertain to certain characteristics of cycling. Cyclists are in direct inter-action with their environment, they do not have a vehicle's bodywork to protect them, and they are propelled by their own muscle energy. While these aspects can act as motivations to cycle, they are also the basis of the barriers encountered by commuters. Four ranges of barriers have been identified: weather conditions, logistical constraints, safety issues and comfort issues.

These obstacles are all different in nature and have varying levels of impact on the practice of cycling. Many are time contingent or linked to specific circumstances and do not prevent cycling in general. This is particularly true of personal barriers (owning a bicycle that meets the cyclist's needs, having appropriate equipment and the required level of physical fitness, riding in bad weather, etc.). Others reflect the territory's hosting potential. More so than climate or topography, safety issues are key. They are more diffuse than the other barriers identified but are also more likely to prevent the adoption or continuation of cycling. They highlight both the problems of coexistence with other road traffic and the need for cycling facilities and infrastructure.

Cycling, and the extent to which it is practised, depends closely on traffic conditions, which themselves depend on urban form, cycling facilities and infrastructure, and coexistence with car traffic, as well as the rules and norms in place. The importance of a territory's level of bikeability—i.e. its potential to hosting a diversity of cycling practices—is key.

A significant level of sensitivity to infrastructure can be observed in the responses, as already highlighted by the literature [1, 3]. Levels of ease are highest on cycle tracks separated from other vehicles, and decrease rapidly depending on the level of cohabitation with road traffic and its speed and volume, elements, which make bicycle users more vulnerable. There are also differences in levels of ease and expertise between beginners and more experienced cyclists, with the former being more sensitive to traffic conditions.

Analysis of home–work journeys shows shortcomings in the bikeability of territories, whether in terms of secure and sheltered parking, continuous cycle routes, or the adaptation of crossroads and intersections for cyclists. Cyclists frequently have to share space with traffic on maladapted roads, and this can be a negative experience. Cohabitation with motorised traffic may be difficult, and the absence of appropriate infrastructures and the lack of legitimacy of cyclists on the roads cause an important minority of cyclists to feel unsafe (one in seven) or disrespected by other road users (one in three). These values are much higher when looking specifically at French-speaking and Italian-speaking Switzerland. This patchy bikeability has various consequences: limiting the practice to certain population groups, restricting it to certain times of the year, adoption of avoidance tactics (longer routes) or breaking traffic rules. A tension arises between cycling as an enjoyable experience and cycling seen as an act of courage.

These results, obtained within a population that actively engages in cycling, show that the current infrastructure in Switzerland suits experienced and convinced cyclists who, due to their skills and motivations, are willing to assume risk because they do not want to give up the benefits of cycling. However, the traffic conditions are not sufficient to encourage cycling more widely as a means of transport. Indeed, the identified problems have an even greater impact on people who are less motivated to cycle or less skilled in the practice; these individuals may be discouraged from adopting utility cycling in the first place, or may give up as a result of the difficulties experienced. In addition, an environment that does not enable the successful coexistence of the different vehicles risks polarising cyclists and non-cyclists in their road use.

Marked differences are observed within the country in terms of the *bike to work* participants' feelings of riding safely and of being respected by other road users. The difference in practice between German-speaking and Latin Switzerland is often interpreted as a consequence of cultural differences, although these differences are not explicit. It appears here to result primarily from differing traffic conditions and political issues around the importance accorded to bicycles and motorised vehicles.

Table A.1 Effects of the *bike to work* event on cycling practice

Motivational effect (group dynamic)	Recruitment (new utility cyclists) Reminder (leisure, sport, seasonal cyclists) Extension (distances, time)
Learning effect (gaining and exchanging experience)	Access to suitable bikes and equipment Skills (physical condition, choice of route, cohabitation with motorists, etc.) Appropriation (feasibility of the commuting trip, etc.)
Legitimising effect (recognition of minority practice)	Sense of belonging A way to 'cast a vote' for cycling

Cities that are renowned for their commitment to promoting the bicycle are charac-terised by a more complete cycling system, a greater modal share for cycling, and a more diversified population of users (as in Basel-Stadt, which is the only canton where women represent the majority).

Bike to Work: **Increasing Commuters' Cycling Mobility Potential**

Bike to work gathers thousands of teams of four employees who commit to cycling to work as much as possible over 1 or 2 months. Several questions from a survey sent to participants, as well as information given in comments interviews, enable us to identify three effects of the *bike to work* action: a motivational effect (a group dynamic that helps recruit utility cyclists), a learning effect (increasing the cycling potential of individuals through the gaining and exchanging of experience) and a legitimising effect (normalising a minority practice) (Table A.1). While many studies focus on the issue of modal shift—which may be the ultimate goal—these results highlight some of the mechanisms that are necessary to recruit new utility cyclists and to convince commuters to change their habits.

The motivational effect stems from the group dynamic induced by *bike to work*. It is a tool of recruitment as it encourages several kinds of people to try and adopt cycling. According to the survey, 1 in 10 participants did not use to cycle to work, and two-thirds claim, 3 months after the end of the initiative, that they have taken up bicycle commuting (this represents several thousand people each year).[1]

The motivational effect creates a period during which the practice of cycling is encouraged. It may lead some car drivers or public transport users to question their habits, kick-starting a change to another means of transport that may see them

---

[1] The survey here is limited due to its methodology: the frequency of cycling to work is not known precisely, and nor is it known whether the practice is continued over the longer term (including winter). Moreover, as stated by some participants, the adoption of cycling may take more than a single participation in bike to work.

becoming utility cyclists. The motivational effect also concerns additional categories of commuters: leisure or sport cyclists may widen their practice (temporarily or not) to utility cycling. This is important, as cycling can actually be divided into several different practices (utility, leisure, sport) that are somewhat separate, at least in the case of Switzerland. *Bike to work* may also be a reminder to seasonal cyclists to start cycling again. Finally, some regular utility cyclists deliberately travel further during the competition than they otherwise would.

The majority of participants already use their bicycles for utility purposes and do so independently of the event. Some temporarily adopt specific practices, such as travelling longer distances. For the participants who did not use a bicycle to get to work prior to *bike to work*, the scheme represents a trigger effect, acting as a recruitment tool that gets people who are interested in utility cycling to try it out. In addition, competition between colleagues can create a group dynamic that may encourage some commuters to get back on their bikes or to extend their cycling practices to include commuting. New cyclists often go on to adopt the practice, even if the effect is not necessarily immediate. For cyclists who typically cycle on a more seasonal or irregular basis, by creating a period during which the practice is encouraged, *bike to work* reminds them to 'get on their bikes'.

The learning effect leads to an increase in cycling potential for neophyte and irregular cyclists. Skills and knowledge are often underestimated in transport studies [11], but our results highlight their crucial importance in the adoption of utility cycling. Even though these skills may appear mundane or self-evident, they require a certain period of learning. This is made possible by the length of time that the Swiss campaign lasts (1 to 2 months, while many others are much shorter) and by the exchanges between regular cyclists and neophytes. As well as improved skills (knowledge of the trip, physical fitness, knowing how to cycle in traffic), the increased individual cycling potential may result from (a) improved access to equipment (through getting their bike back into working order, or buying a bike that is better adapted to their needs, for example) and accessories (waterproof clothing, bike locks, etc.) and/or (b) appropriation (feasibility and attractiveness of bicycle commuting).

These first two effects demonstrate that cycling has a dynamic aspect. An individual's cycling practice inevitably changes over time, and these potential changes should be taken into account when promoting the practice. These may include varying continuity of practice, abandonment (following a move, a new job, etc.) or resumption after certain events (such as participation in *bike to work*).

The legitimising effect makes utility cycling more visible and creates a mass effect.[2] It helps to normalise and legitimise cycling, which is still a minority practice, particularly in the professional world, where cars are often strongly anchored within travel habits. *Bike to work* contributes to a sense of belonging and a sense of community. This kind of argument gradually gains in importance with the number of participations in *bike to work*.

---

[2]One initiative that helps to legitimise cycling is Critical Mass, which involves large-group bike rides occurring in cities around the world, usually on the last Friday of every month. This helps to make cycling more visible and to reclaim space back from motorised traffic.

For novice cyclists whose perception of their cycled commute has changed, the change is generally positive with regards to duration, effort and how pleasurable it is. The effect is much less positive, however, in terms of safety, which points to shortcomings in cycling facilities and infrastructures in Switzerland. This observation underscores one of the limitations of the scheme: it does not impact on the hosting potential of the territories in terms of infrastructures. In other words, while *bike to work* represents a favourable period for the practice of cycling, it comes up against restrictions associated with the territory and its organisation.

The Need for Legitimisation and Safe Infrastructure in Order to Promote Cycling

Half of the *bike to work* participants have a negative opinion of public authority commitment to cycling in their region. Here too, the differences between regions and cantons are very marked. People are most critical of the public authorities in the French- and Italian-speaking cantons.

The measures recommended by cyclists in respect of the bikeability of their home–work commute focus on two dimensions: first, bicycle urbanism (the *hardware*, material artefacts) and second, the development of norms and rules that take cycling into account and manage traffic conditions (the *software*). Measures to improve the bikeability of territories and increase their hosting potential for cycling would also, in turn, promote and encourage cycling. It is also possible to formulate guidelines for promoting cycling using our theoretical framework of velomobility and by focusing on the three dimensions of mobility defined by Cresswell: movement, experience and meaning [5, 6].

Promoting cycling begins with improving the *movement*, from point A to point B of the journey. It involves offering fast, direct and networked routes, and this means cycling infrastructure (cycle tracks or even lanes, bicycle expressways, advanced stop lines, maintenance and equipment of routes, etc.), favourable parking conditions, complementarity with public transport, certain traffic management rules (contra-flow cycling, synchronisation of lights with cycling speed, turn-right-on-red policies, limiting the speed of motor vehicles in certain places, etc.) and an urban form based on proximity, density and a diversity of functions (housing, economic activities, leisure, etc.).

The promotion of cycling also requires that the *experience* of the practice be shown due consideration. This means providing suitable, safe, but also pleasant and attractive, routes for a large number of bicycle users, and includes planning measures as well as the evaluation of good practice. It also involves a more contextualised and precise knowledge base, as well as greater sensitivity—on the part of urban planners, transport engineers and other professionals in the field—with regards to the needs, practices and feelings of the different types of cyclist.

There is a third dimension underlying those of the movement and experience of the cycled journey: its *meaning*, which has two levels. At the societal level, it is a question of legitimising the position of cycling within the mobility system, of giving it a place within a territorial context, which is still dominated by cars and of considering it to be a means of transport in its own right. At the individual level (that of the cyclist, whether established or potential), communication and promotional campaigns should be developed around positive and unifying images and representations of cycling.

These measures primarily concern the hosting potential of the territories under study. The preponderance of comments of this nature made by the *bike to work* participants is explained by the fact that the survey related to their commute and that many of them have already overcome their own personal barriers. Other measures, relating to individual cycling mobility potential—access, skills and appropriation— should be considered in the context of an integrated cycling policy.

Access concerns, first of all, equipment. Measures are required that promote access to a bicycle that meets the user's needs (whether this is mechanical or electrically assisted, owned or shared, foldable or cargo, etc.). This may include, for example, grants to purchase second-hand bicycles, bicycle sharing schemes, subsidies, better tax recognition of commuting trips made by bicycle, etc. In addition, good bicycle maintenance is important in order to ensure the reliability of travel by bike. Maintenance services can be provided by bicycle shops, or skills can be taught through participatory workshops.

In terms of skills, initial training must promote the renewal of the practice, within a context where both children and adolescents cycle less than in previous generations. Other courses could supplement this offer periodically, targeted at adults who don't know how to ride a bicycle or no longer do so, people who want to use other types of bicycle (electrically assisted, cargo, etc.) or cyclists who wish to improve their level of ease, especially in traffic. Specific services can offer assistance and support in the field, such as signage, maps, guides or dedicated apps.

Finally, with regards to appropriation, several measures act as an incentive or reminder to cycle. These include similar events to *bike to work* (e.g. *bike2school*, or *Défi vélo* for high schools in Switzerland) or car-free days, as well as communication campaigns setting forth positive and targeted messages around the benefits—both personal and societal—of cycling. The challenge here is, first, to encourage people to go beyond the inertia of their regular mobility practices and, second, to transform occasional behaviours—such as leisure cycling on holiday or in summer—into habits, in order to turn cycling into a normalised means of transport.

Towards a Coherent System of Velomobility

Debates around the future of transportation are dominated by promises of technological solutions that ensure an accessible, efficient and clean mobility. However, given the magnitude of the challenges, a plurality of responses and combinations of responses will be required. The different forms of active mobility and, in particular, cycling will form part of this approach. Silent, resource-efficient, inexpensive, user-friendly, and sustainable, bicycles have a lot of advantages.

The example of Dutch cities and Copenhagen was often mentioned in the comments left in the questionnaire. They feed the imagination; they are benchmarks and a source of inspiration for bicycle urbanism. However, it should not be forgotten that these cities, like the rest of the continent, experienced a collapse in utility cycling in favour of cars and motorised two-wheelers until the 1970s. Their current situation owes nothing to chance, nor to any particular geographical context or cultural predispositions. First and foremost, it is the result of more than 40 years of integrated and coherent cycling promotion policies.

In Switzerland, utility cycling faces a certain number of barriers, and cycling culture is not as developed as in northern Europe. There are, however, signs of a cycling renaissance. Utility cycling is gaining popularity in cities and also has a presence within other spatial contexts. Real-cycling infrastructures are emerging. New types of bicycles and services enable us to consider additional uses and to reach new audiences, and some communities are adopting relatively ambitious plans that offer a glimpse of a more comprehensive bicycle system.

However, the pitfalls are still numerous. The analyses at the heart of this work echo the scarcity of amenities, decry infrastructure designed without taking into account the specificities of active mobility, call out the lack of continuous, coherent cycle routes and highlight the challenge of cohabitation with car traffic. For public authorities claiming, they want to promote active mobilities and respond to the various societal and environmental challenges of the transportation system, it is important that they recognise cycling as a means of transport in its own right and integrate it into the transport system. Providing a physical and social environment that is conducive to cycling and a coherent system of velomobility is a matter of consistency and a proof-in-action of commitment to a transition towards a more sustainable mobility.

References

1. R. Aldred, B. Elliott, J. Woodcock, A. Goodman, Cycling provision separated from motor traffic: A systematic review exploring whether stated preferences vary by gender and age. Trans. Rev. **37**(1), 29–55 (2017). https://doi.org/10.1080/01441647.2016.1200156
2. F. Behrendt, Why cycling matters for electric mobility: towards diverse, active and sustainable e-mobilities. Mobilities **13**(1), 64–80 (2018). https://doi.org/10.1080/17450101.2017.1335463
3. R. Buehler, J. Dill, Bikeway networks: a review of effects on cycling. Trans. Rev. **36**(1), 9–27 (2016). https://doi.org/10.1080/01441647.2015.1069908
4. S. Cairns, F. Behrendt, D. Raffo, C. Beaumont, C. Kiefer, Electrically-assisted bikes: potential impacts on travel behaviour. Trans. Res. Part A Policy Prac. **103**, 327–342 (2017). https://doi.org/10.1016/j.tra.2017.03.007
5. T. Cresswell, *On the Move: Mobility in the Modern Western World* (Routledge, London, 2006)
6. T. Cresswell, Towards a politics of mobility. Env. Plan. D Soc. Space **28**(1), 17–31 (2010). https://doi.org/10.1068/d11407
7. E. Fishman, C. Cherry, E-bikes in the mainstream: reviewing a decade of research. Trans. Rev. **36**(1), 72–91 (2016). https://doi.org/10.1080/01441647.2015.1069907

8. S. Handy, B. van Wee, M. Kroesen, Promoting cycling for transport: research needs and challenges. Trans. Rev. **34**(1), 4–24 (2014). https://doi.org/10.1080/01441647.2013.860204
9. B. Ingels, Hedonistic sustainability (2011). https://www.ted.com/talks/bjarke_ingels_hedoni stic_sustainability. Accessed 01 August 2020
10. H. Jones, K. Chatterjee, S. Gray, A biographical approach to studying individual change and continuity in walking and cycling over the life course. J. Trans. Health **1**(3), 182–189 (2014). https://doi.org/10.1016/j.jth.2014.07.004
11. V. Kaufmann, *Rethinking the city: urban dynamics and motility* (Routledge &; EPFL Press, Lausanne, 2011)
12. V. Kaufmann, M.M. Bergman, D. Joye, Motility: mobility as capital. Int. J. Urban Reg. Res. **28**(4), 745–756 (2004)
13. V. Kaufmann, E. Ravalet, E. Dupuit (eds.), *Motilité et mobilité: mode d'emploi* (Éditions Alphil-Presses universitaires suisses, Neuchâtel, 2015)
14. J. MacArthur, J. Dill, M. Person, Electric bikes in North America: results of an online survey. Trans. Res. Record J. Trans. Res. Board **2468**(1), 123–130 (2014). https://doi.org/10.3141/246 8-14
15. D. Marincek, P. Rérat, From conventional to electrically-assisted cycling. A biographical approach to the adoption of the E-Bike. Int. J. Sustain. Trans. (2020). https://doi.org/10.1080/15568318.2020.1799119
16. D. Piatkowski, R. Bronson, W. Marshall, K.J. Krizek, Measuring the impacts of bike-to-work day events and identifying barriers to increased commuter cycling. J. Urban Plan. Devel. **141**(4), 04014034 (2015). https://doi.org/10.1061/(ASCE)UP.1943-5444.0000239
17. J.R. Pucher, R. Buehler (eds.), *City Cycling* (MIT Press, Cambridge, 2012)
18. D. Sauter, K. Wyss, *Etude pilote sur l'utilisation du vélo chez les jeunes dans le canton de Bâle-Ville* (Département des constructions et des transports & OFROU, Bâle & Berne, 2014)
19. P. Walker, *Bike Nation: How Cycling Can Save the World* (Yellow Jersey Press, London, 2017)
20. A. Wolf, S. Seebauer, Technology adoption of electric bicycles: A survey among early adopters. Trans. Res. Part A Policy Pract. **69**, 196–211 (2014). https://doi.org/10.1016/j.tra.2014.08.007